THE UNITED NATIONS UNIVERSITY/THIRD WORLD FORUM

STUDIES IN AFRICAN POLITICAL ECONOMY

The Mediterranean

THE UNITED NATIONS UNIVERSITY/THIRD WORLD FORUM

STUDIES IN AFRICAN POLITICAL ECONOMY
General Editor: Samir Amin

The United Nations University's Project on Transnationalization or Nation-Building in Africa (1982–1986) was undertaken by a network of African scholars under the co-ordination of Samir Amin. The purpose of the Project was to study the possibilities of and constraints on national autocentric development of African countries in the context of the world-system into which they have been integrated. Since the 1970s the world-system has been in a crisis of a severity and complexity unprecedented since the end of the Second World War; the Project examines the impact of this contemporary crisis on the political, economic and cultural situation of Africa today. Focusing on the complex relationship between transnationalization (namely, the dynamics of the world-system) and nation-building, which is seen as a precondition for national development, the Project explores a wide range of problems besetting Africa today and outlines possible alternatives to the prevailing development models which have proved to be inadequate.

TITLES IN THIS SERIES

M. L. Gakou
The Crisis in African Agriculture
1987

Peter Anyang' Nyong'o (editor)
Popular Struggles for Democracy in Africa
1987

Samir Amin, Derrick Chitala, Ibbo Mandaza (editors)
SADCC: Prospects for Disengagement and Development in Southern Africa
1987

Faysal Yachir
The World Steel Industry Today
1988

Faysal Yachir
Mining in Africa Today: Strategies and Prospects
1988

Faysal Yachir
The Mediterranean: Between Autonomy and Dependency
1989

Azzam Mahjoub (editor)
Adjustment or Delinking? The African Experience
1989

Hamid Ait Amara, Bernard Founou-Tchouigoua (editors)
African Agriculture: The Critical Choices
1989

Other titles in preparation.

THE UNITED NATIONS UNIVERSITY/THIRD WORLD FORUM

STUDIES IN AFRICAN POLITICAL ECONOMY

The Mediterranean

Between Autonomy and Dependency

Faysal Yachir

The United Nations University
Tokyo
Zed Books Ltd.
London and New Jersey

The Mediterranean: Between Autonomy and Dependency
was first published in 1989
by:
Zed Books Ltd., 57 Caledonian Road, London N1 9BU, UK, and
171 First Avenue, Atlantic Highlands, New Jersey 07716, USA
and:
The United Nations University, Toho Seimei Building,
15–1 Shibuya 2-chome, Shibuya-ku, Tokyo 150, Japan
in co-operation with
The Third World Forum, B.P. 3501, Dakar, Senegal.

Cover designed by Andrew Corbett.
Typeset by EMS Photosetters, Rochford, Essex.
Printed and bound in the United Kingdom
at Bookcraft (Bath) Ltd.

British Library Cataloguing in Publication Data

Yachir, Faysal
The Mediterranean
1. Mediterranean region. Economic
development
I. Title
330.9'182'2

ISBN 0-86232-858-6
ISBN 0-86232-859-4 pbk

Library of Congress Cataloging-in-Publication Data

Yachir, F.
The Mediterranean between autonomy and dependency/Faysal Yachir; with an introduction by Samir Amir.
p. cm. – (Studies in African political economy)
Includes index.
ISBN 0-86232-858-6.–ISBN 0-86232-859-4 (pbk.)
1. Mediterranean Region – Economic integration.
2. Europe, Southern–Foreign economic relations–Africa, North. 3. Africa, North–Foreign economic relations–Europe, Southern.
I. Title.
II. Series.
HF1582.7.Y33 1989 89-8872
337.1'1822–dc20 CIP

Contents

Acknowledgements

This book was produced in the framework of the United Nations University 'African Regional Perspectives' programme under the auspices of the Third World Forum. We wish to thank the UNU, the Italian Cooperation Agency and the Swedish Agency SAREC whose financial support made this work possible. Nevertheless, in accordance with the usual formula, the opinions expressed here are those of the authors alone and not of those institutions.

Preface

From earliest antiquity to the 15th century, the Mediterranean was the centre of that part of the ancient world situated to the west of India and China. It was here, around the eastern basin of the Sea (Egypt, Babylonia, Persia, Phoenicia and Greece) that the earliest civilizations appeared. From the time of Alexander the Great's conquests, Hellenism gave the region a common imprint. Then, within the framework of unification under Rome, this civilization was gradually transferred to the peoples of North Africa and north-west Europe. Thus the foundations were laid upon which the medieval worlds of Christendom and Islam were later built. For 1,000 years there was a constellation of societies bound to one another by organic ties of culture and ideology and technological and commercial exchanges to such an extent that it became possible to speak of a system. Some of the key features of capitalism (commodity exchange and merchant capital, free wage labour, private ownership of the land and enterprise) appeared in the region at an early date and, at some periods – notably during the early centuries of Islam and at the time of the expansion of the Italian cities (between the 12th and 14th centuries) – became distinguishing features of segments of this system, so much so that the 'Mediterranean system' can be seen as the prehistoric ancestor of the world-system of modern capitalism.

The Renaissance marked a qualitative break with this past because from then on the scattered elements of proto-capitalism crystallized to give birth to a coherent new social system: capitalism. When that happened, the relationship between power and wealth was at once reversed. Previously, wealth had, in the last analysis, always been dependent on power, but with the Renaissance it was economic wealth that defined the content of political power. By the same token, the old metaphysical ideological constructs (of Hellenism, Christianity and Islam) that were consonant with the requirements of a system based on the tributary mode of production, gave way to a new political construct and a new type of universalist aspiration. At the same time, the Renaissance saw the centre of gravity of the new capitalist world shift from the shores of the Mediterranean to those of the Atlantic. The old periphery of the Mediterranean system – north-west Europe – became the centre of the new European and Atlantic capitalist world-system.[1]

The Mediterranean region, in turn, was transformed into a periphery as the capitalist system developed. Later, the southern Arab shore was even colonized, while the late formation of the bourgeois national state in Italy and the Balkans meant that obvious traces of underdevelopment persisted. The Mediterranean ceased to belong to the people living along its shores and became a geostrategic region for others, dominated by the hegemonic power, first Great Britain, then the United States, or fought for by those challenging it: Germany and later the USSR.

Have things changed, or are they in the process of doing so? After the Second World War, the European Mediterranean countries, except for Yugoslavia and Albania, were caught up in Western reconstruction under the wing of the United States and then gradually integrated into an EEC broadly subjected to the dominant forces of global transnationalization. And while they took off in terms of economic growth, their development completed the process of tying their future to that of their European partners and, beyond that, to developments in the advanced capitalist centres as a whole. The Arab states' attempt to reconstruct themselves as national bourgeois states, has so far been without success. The illusions of the 'Bandung project' (1955–73), defined by the objective of an autonomous modernization in the framework of global interdependence,[2] followed by those of the oil boom (1973–82), were succeeded by the still continuing period of recompradorization. The comparison with Turkey is obvious here: after the collapse of the Ottoman empire, Kemalism had nurtured the same project, of which the least that can be said is that it has not succeeded either in 'Europeanizing' Turkey or in eliminating the character of a peripheral capitalist society.

This twofold evolution intensified the division between the two sides of the Mediterranean to the extent that it became the frontier of North–South confrontation. In these circumstances, various possibilities are open. Either popular social forces will impose the reconstruction of the Arab world as a united whole, in the framework of a strategy delinked by force of circumstance from the logic of the global expansion of transnationalized capital. In the most favourable hypothesis this reconstruction might be part of a global peaceful evolution towards a polycentric world. For that Europe would have to distance itself from its Atlantic ties and look favourably on the Arab revival. Or the current misunderstandings will continue and the confrontations will get worse. The Europeans might then well pursue the chimerical project of an imperialist revival, and endeavour to bind the Maghreb to their side – like Turkey – while Egypt and the eastern Arab world would be handed over to the regional hegemony of the Zionist state.

This book pursues this argument, avoiding detailed descriptions of events in each country in the region. Samir Amin's introduction sets out, in the light of the historical background and current confrontations, the conditions for a harmonious and peaceful development for the benefit of the peoples of the region, against the background of the construction of a polycentric world. In the chapters that follow, Fayçal Yachir analyses the scale of the contrast that has emerged as a result of the divergent economic paths taken since the Second World War north and south of the Mediterranean.

Notes

1. This argument is developed in: Samir Amin, *L'eurocentrisme, critique d'une idéologie*, (Zed Books Ltd, 1989). The argument stresses the features common to the various regions of the world descended from Hellenism (Christian Europe and the Arab–Islamic world) and places the significance of the break at the time of the Renaissance in this context. Fragments of this argument have been published in English as 'Some reflections on Arabo-Islamic thought', *Bulletin of the Third World Forum* (Dakar), No 7, 1987.
2. See Samir Amin, *Bandung, trente ans après*, UNU, Cairo 1985.

1. Conditions for Autonomy in the Mediterranean Region*

Samir Amin

The Weight of the Past

The Mediterranean area, by virtue of its past and many of its present characteristics, is evidently a region where diversity does not preclude unity, and both are products of geography and history. The region is also evidently not cut off from the rest of the world. On the contrary it has become one of the key areas of confrontation and conflict in our time. It is an area subject to contradictory forces, some tending towards homogenizing assimilation to the rest of the world and others towards the ravages that can ensue from the unequal character of the development of capitalism or from rejection of capitalism.

No more than elsewhere is the history of development of the Mediterranean region linear. One can point to two successive main periods that are broadly divergent in their effects.

From earliest antiquity until the late Middle Ages and the Renaissance, the main determining factor was the deepening of intra-Mediterranean relations, that despite their conflicting nature emphasized the region's unity. Here civilization begins very early, first on the banks of the Nile, then slowly diffused along the eastern shore of the sea, then along all – or almost all – its shores. From being mainland and regional the (Egyptian) civilization turned towards the sea, and with the Phoenicians and Greeks took on a sea-going character, that created the conditions for unity of the region.

The Mediterranean region gradually became the centre of the ancient world to the west of India (further to the East, in India and China, other centres were established that had scant relations with the powerful western Mediterranean world). The region had a central position unthreatened from outside. Europe to the north of the Mediterranean shores and Africa to the south of the Sahara

* This contribution to the collective work draws extensively on two previous texts: *Les Conditions d'une Autonomie de la Région Méditerranéenne* (Naples Symposium, 1983); *Le Contexte Economique des Relations Euro-Arabes* (Mons Symposium, 1984).

We also take up several theses developed in greater detail elsewhere (see Notes 4, 6 and 7).

were neither a challenge, nor even a barbaric threat – the other regions were too backward then in comparison with the Mediterranean area.

Only on its undefined eastern mainland borders was the region open to the movements of the Asian populations. But here, too, the main movements were in directions that did not threaten the Mediterranean. Two other autonomous centres were in fact constituted in the still remote Asia: the Chinese centre and the Indian centre. But the Han and their civilization spread from the north of China towards the south (South China, Vietnam) and advanced towards the east (Korea, Japan), and crowded over a vast territory; but they were isolated from the West by the barrier of the Himalayas and the central Asian deserts. As for Indian civilization, it extended eastwards down the Ganges to merge into the sub-continent with the Deccan peoples, without a glance westwards. The only Mediterranean window open was to the gusts of Central Asia, whose populations, blocked to the east and south-east by the Chinese and Indian masses, were always drawn westwards, from the adventures of ancient Persia to the Turkish empire of the Muslim Khalifate. Reasons of defence and the legend of an Orient rolling in wealth occasionally led Mediterranean figures, such as Alexander the Great, to venture far eastwards of their home base and in the end to become lost without being able to do more than touch on India and China.

With these uncertainties, the Mediterranean, as centre of this part of the world, acquired a unity that overrode its diversities.

There is no easy or obvious explanation of such Mediterranean unity. The anthropologist who is drawn to detail is struck by the diversity. Geography, unless one is going right back to the late Stone Age, points to diversity rather than unity. Obviously one cannot talk in the same way of the crowded societies of Egypt, irrigated for the past five thousand years, those of the sparse, non-irrigated, agricultural areas that predominate to the north of the Mediterranean, and those of nomadic pastoralism in the Middle East and North Africa. Political unity – of the imperial kind – was achieved only once, in the brief period of the Roman Empire. Neither Byzantium, nor Arabian Islam, nor the Ottoman Empire succeeded in restoring it. Islam itself was to remain torn between the western pole of attraction – illustrated by the advance of the Ottomans as far as the gates of Vienna – and expansion east and south, that brought its major success from the Indian sub-continent to Indonesia and the sub-Sahara from Dakar to the Bight of Benin and to the Indian Ocean coast. But the absence of a unifying imperial power did not impede intensive exchanges of every kind that the inland sea required. Hence an urban-type civilization prevailed in a Mediterranean that cannot be explained without the trading functions of the town, despite the variety of social formations at the base that were necessarily rural, as everywhere else until modern times, despite ethnic and cultural irredentism, despite even the great religious cleavages (Rome/Byzantium, Islam/Christendom). This urban character is in contrast to rural dominance in northern, central and eastern Europe. This character is also the source of other Mediterranean affinities: there was never any kind of feudalism, as there was in the rest of Europe, but rather the power of merchant and clerical urban aristocracies. The urban character has left its mark on modern politics; and it was doubtless not unfamiliar with the phenomenon of clienteles.

The thesis we have developed elsewhere on unequal development in the pre-capitalist era and the birth of capitalism is based precisely on the contrast between the lead of the (Italian and Arab) Mediterranean in this first centre – which was later to be a handicap – and the backwardness of the European feudal periphery – which would be an advantage in the birth of capitalism. That is how we explain the 'shift of the centre of gravity' from the western world of the Mediterranean to the Atlantic, without attaching as much significance as others to chance (the crossing of the Atlantic and discovery of America), but on the contrary to the deepening of the concept of unequal development and the supersession of a system from its peripheries.[1]

Nevertheless, until this supersession – from the sixteenth century – the Mediterranean really was the centre of this part of the world. It owed little to external exchanges and much to the relations between its constituent parts. Hence the hegemonies in the region are as follows: the successive Mediterranean worlds of the Phoenicians, Greeks, Imperial Romans, Arabs (in their initial grandeur of the first three centuries of the Hegira), the Italian cities, and the Ottoman Empire.

Traces of this first great epoch – of the middle-earth and world-centre Mediterranean – are still visible. But it is dangerous to jump from this observation to talk of the 'Mediterranean community', which smacks too much of geographical pseudo-determinism and 'peoples' psychology'. For this first epoch is over and done with.

During the following four centuries, in the course of which the world we now know was shaped, the main movement was in the other direction: shattering the Mediterranean region through centrifugal attractions of a world that was no longer tied to its shores. To be more exact, the Mediterranean became the frontier zone between the new centre – European then Euro-North-American – and the new periphery – Afro-Asian.

This fracture emphasized the oppositions between the new crystallizations. Europeanism called the tune, since it was associated with the formation of the new capitalist and European centre, although it was henceforth impossible to separate the two aspects of the one reality.[2] An avatar of Christendom? The creed, of Mediterranean and Oriental – not to say Egyptian – origin, spread into the barbarian North where it flourished, while to the south of the inland sea it faded out and gave place to Islam. The new reality of Europe seeks its supposed roots and ideological justifications in the ancient Mediterranean world that nurtured it: from the Renaissance rediscovering Greece and Rome to contemporary talk in EEC Europe, making Athens the cultural capital of Europe, there is no shortage of such a quest for origins. But it is interesting to note here that these supposed roots are sought exclusively in the regions of the Mediterranean area that have remained Christian. Recognition of the role of Egypt and Islam is left to rare specialists; an appeal to popular feeling here would be regarded as almost indecent.

The crystallization of the Arab nation was a product of reaction to the new challenge, nothing to do with the challenges of the previous centuries, even allowing for that of the Crusades. The Arabization and Islamization from the Atlantic to the Gulf are undoubtedly earlier and so an Arab nation was fully in existence in the first centuries of Islam, then it its first glory. Evidence, too, of this region's lead over feudal and fragmented Europe: the centralization of surplus by the class of warrior merchants, the alliance of the cities they led and the Khalifate, the support of the nomads, in order to keep control of communications and the countryside, are the foundations of this nation.[3] Yet it later decayed, with the decline of the great trade and the call for the help of the Turkish barbarians of Central Asia. The Ottoman reunification did not halt the process, but even to some extent accelerated it. Hence the renaissance of the Arab nation would come in dual reaction to the European challenge and Ottoman domination. This renaissance began early, since the threat of European advance was quickly felt, in comparison with the Indian, Chinese and black African worlds, which were much less well-known to Europe until a later date and worlds that often knew nothing of Europe except through the Arabs. The consciousness of the gap came quickly – at least in Egypt – as is shown by the efforts of Ali Bey El Kebir in the eighteenth century, that is only a century or so after the gap first came into being. The unrelenting hostility of Europe to Mohamed Ali's attempt to modernize the Nile Valley, to raise the dignity of and free the Arab Mashreq (in the first half of the nineteenth century) has become a constant feature of the West's strategy towards Egypt. The hegemonic powers of the capitalist centre – Britain in the nineteenth century, the United States since the Second World War – have always deemed it essential to their predominance to maintain Egypt in such a ruinous condition that it could not become the pivot of a revived Arab nation, that is a genuine partner in the worldwide capitalist system. The plan of creating an artificial European state in Palestine, to undermine such a possibility, was dreamed up by Palmerston twenty years before Zionism even took shape.

Nonetheless, this gradual two-fold crystallization created a divide that had not existed in earlier times. Until then Euro-Arab relations, to use modern terms, were characterized by a kind of 'conniving rivalry' between equals, who could therefore play out their own internal contradictions without fear or shame, as the events of the Crusades, among others, have shown. By contrast, the divide that capitalism opened up between Europe and the Arab world changed the basis of such relations. Doubtless, capitalist development was unequal even within Europe, and also opened up another divide between Northern Europe and Southern Europe, which had been in the forefront in earlier times and declined gradually to the level of what is sometimes described as semi-periphery; but this does not prevent Southern Europe being part and parcel of Europe. Moreover, when capitalism began, Spain and Portugal, after the decline of the Italian cities, founded the first mercantilist empires by conquest of America, before the hegemony in dispute between France and England for two centuries passed definitively to the latter (with the Treaty of Paris of 1763 granting India to Britain). Portugal, albeit impoverished, remained a colonial power, in Africa at least, and waged colonial war until 1974, an indication of the extent to which that country considered itself part of the imperialist world despite everything and even though solely in a subordinate intermediary role. Italy too dreamed of a 'Mare Nostrum' and tried with its paltry means to put flesh and blood on this dream. France, a significant imperialist power, provided itself with a southern Mediterranean shore and until recently dominated the western half of the Arab world.

Did not colonization, a recent (nineteenth century) phenomenon, open a definitive divide and turn the Mediterranean into a frontier zone of the main confrontation of our time: between North and South? For colonization wrought inequalities of economic development considerably more reprehensible than in the past, difficult to reverse except by recourse to a perspective diametrically opposite to that of the expansion of the world capitalist system from its outset. Colonization has also revealed a moral and political contrast, and given the religious dimension (of Christianity and Islam) a weight it did not have in the past and one now capable of nurturing fanaticism.

It is clearly understood that as the hegemonic centres of the worldwide capitalist system lie outside the Mediterranean region, the Sea ceases to be the centre of its world to become a geo-strategic zone for others. From the destruction of Napoleon's fleet at Trafalgar until 1945, Britain dominated the Mediterranean – which provided her shortest route to India – and occupied Egypt, Cyprus, Malta, Gibraltar and Palestine; its fleet had no rivals. Neither France, despite all its trump cards – which became Britain's ally – nor Germany – which played the Ottoman card, then the Axis card in alliance with Italy – nor Russia – handicapped by the Straits Convention (of 1841) – could succeed in their attempts to challenge the 'British Mediterranean'. This was reluctantly ceded, after the Second World War, to give way to the era of the 'American Mediterranean' in which we still find ourselves.

Post-war: from the great upsurge of capitalist expansion to the contemporary crisis

We shall not repeat what we have written elsewhere about the principal characteristics of the great upsurge of capitalist expansion between 1945 to 1970, following the Second World War: worldwide penetration that reaches a qualitatively new stage, with the internationalization of imperialism, and American hegemony that provides the framework for this almost unprecedented upsurge. For the Mediterranean region in particular, the period is one of European construction (with an impact on Southern Europe) and of development of the Arab plan for unitary and popular liberation (and its confrontation with the Palestinian issue). We shall also say no more on the character of the crisis that brought an end to this upsurge, about which we have commented elsewhere. It is necessary, however, briefly to recall our general conception, for this conception dictates our analysis of North–South relations in general, and intra-Mediterranean in particular.[4]

Our thesis is that in the current crisis, a long structural crisis of the capitalist system, the organization of North–South relations and the conflict around these relations is the central axis. This crisis is certainly not the first. The entire history of capitalism is made up of such long cycles and the latest long cycle is constituted by the rise of imperialism from the end of the nineteenth century to the 1914 war, followed by a long period of crisis from 1914 to 1945 marked by two world wars and the crisis of the 1930s. We have thus had a period of 25 years, from 1945 to 1970, of strong growth within the system of American hegemony. And since 1970 the start of a long crisis whose end and resumption is announced every quarter, but one which in our view threatens to continue into the twenty-first century if it does not end in the general catastrophe of a huge world war. It is very difficult to know how the world will look at the end of this long crisis. There is a broad variety of explanations for the crisis. Is it a matter of deficient demand? Is it on the contrary a fall in the rate of profit due to loss of flexibility in the system, stemming precisely from the stiffening of the system in the previous period of prosperity?

Let us simply say that the crisis of 1930 came after a series of defeats of the working classes in the West – when the Russian Revolution did not, as Lenin and the Bolshevik Party expected, spread throughout Europe. In these circumstances, the redistribution of income through the Keynesian policies of the 1930s might have been a solution to the crisis. By contrast, our crisis comes after a long period of full employment, the rule of the social democratic welfare state, accepted even by governments of the right, with wages rising along with the rate of labour productivity. That is why North–South relations constitute the central axis of the current crisis. Since if there is deficient demand, it is essentially deficient demand from the South. The cause is super-exploitation in the newly industrialized countries (including Arab countries) characterized by wages that for the same productivity are markedly lower than those in the West. It is, therefore, deficient demand from the South and its deterioration in the previous growth period that originates and characterizes this crisis. A redistribution of demand at global level to permit a fresh start would entail very profound modifications in the international division of labour, especially between Europe and the Arab world. The obvious question is 'under whose aegis' will this reconversion of the international division of labour be carried out, and will it take the form of redeployment of the capital of the American and European imperialist monopolies, or will it rather be carried out under the direction of national powers with popular backing? Will it take place in the various Arab countries or in an Arab nation that is far from being unified but whose eventual unification is a historical necessity?

But it is also a crisis for American hegemony whose decline has begun, since post-war expansion has restored to Europe and Japan a competitiveness they had lost in the course of the Second World War. In these circumstances will Europe revert to being an autonomous imperialist force in regard to the United States? A close look at this highly complex issue will shed light on the ambiguities of Europe's relations with the Third World, the United States and the USSR. The complexity of the question stems from the entanglement of several factors we shall shortly examine.

European construction, initiated with the Marshall Plan and reaching a major turning point with the Treaty of Rome in 1958, is a far from complete or irreversible commitment, despite what the majority of European political forces and populations probably believe. It has only begun its extension to the poorer, Southern Europe (Spain, Portugal and Greece), contributing to a worsening of the contradictions of interest between Southern Europe and the wealthier Europe, and in a period of crisis. The post-war upsurge has already had a substantial impact on the givens of the North–South issue in Europe. For the upsurge has, with its wider world impact, entailed a substantial speeding up of the modernization of the European peripheries: Italy, Spain and Greece. To the point that our colleague Giovanni Arrighi[5] supports the thesis (which we share) that the states of Southern Europe are now so integrated into the European and world system that it is virtually impossible for the dominant political forces of these countries – either of the right or the left – to envisage a response to the crisis by a withdrawal into themselves. There is a striking illustration of this change in the contrast between the attitudes of these countries in response to the crisis of the 1930s – a semi-autarkical withdrawal of populist or fascistic bent – and the current belief of the right and of the left, that acceptance of the rules of the game of worldwide competition is 'unstoppable'.

Nevertheless, European construction remains ambiguous in meaning and prospects; and the challenges of the current crisis, far from attenuating these ambiguities, serve rather to reveal the irresolute and indecisive attitudes of the European partners.

From the outset the European construction had been conceived as a necessary venture to avoid the spectre of 'communism' that has today totally disappeared, if it ever really existed. In this sense, it was conceived as an integral part of the economic, political, military and ideological strategy of United States' domination. European economic integration, far from aiming at the creation of a new autonomous pole competing with the United States, was conceived as a sub-set of the worldwide whole. Europe was open to the Atlantic military alliance and the penetration of the American transnationals who have played a decisive role in its economic modernization. It remains so. First, since it remains under the supposed protection of the United States nuclear umbrella and has not developed an autonomous defence, in the absence of which an autonomous economy is inconceivable. With a touch of bizarre economic shortsightedness, it has been suggested that savings on military expenditure will allow a better economic performance . . . The intention of autonomy, which De Gaulle obviously favoured, never went beyond the stage of irresolute actions, and despite French talk of autonomous defence – almost entirely unanimous talk but for internal electoral consumption – no strategist doubts the narrow limits of the 'strike force'. Furthermore, in response to the challenges of the crisis, Europe has rallied behind the United States in a common Western offensive intended to 'recompradorize' the Third World. This final ambiguity leads us to the issue of imperialism in general, and of

European imperialism towards the Arab and African worlds in particular. Great Britain and France had virtually shared out the Arab and African world between them; and on the morrow of the last world war they did not yet suspect that they would have to bow to the decolonization that was imposed upon them by the liberation movement and acceptable under certain conditions to American hegemony. Decolonization did not come about without conflict; the Algerian war was colonialism's death throe.

European construction had prepared nothing in this regard, except to put the former French colonial empire in North Africa at the disposal of the capital of the Community of the Six, with collective neo-colonialism replacing the former imperial colonialism and little more. Without taking up time with issues that are treated elsewhere, it is necessary to recall our principal theses on this complex subject.[6] First, that France has *de facto* retained privileged status in its former colonies, notably by the bias of control over the franc zone. Second, that the conventions of the association of the African states to the EEC (the Yaoundé Convention then the Lomé Conventions) show a concern for reserving privileged status for Europe in regard to American and Japanese competitors, despite the general opening-up of Africa implied by worldwide expansion. Third, that with Britain's membership of the EEC, and the association of the African states, jockeying for influence by the various powers in Africa is even more overt. Fourth, that the kind of unequal relations renewed in this framework in no sense represent progress towards the liberation of Africa and the development of its peoples; on the contrary, their restriction to obsolete mining and agricultural specializations is to Europe's advantage. In that sense Europe bears a heavy responsibility for the crystallization of the power of the new local ruling classes and thereby for the continent's economic, social and political disaster.

The European view of the Arab world, especially North Africa, scarcely goes any further, except that it had to take into account the greater stability of the local ruling classes. The association agreements drawn up with Morocco and Tunisia made do with providing preferential and provisional access to the European market for the agricultural exports of these countries (until the integration of Southern Europe into the EEC provoked a crisis for these exports), and with relocation (also provisional) of labour-intensive (mainly textile) industries directed towards European exports (until the current crisis called these concessions into question).[7] The strategic view implicit here plunged the Arab partners deeper in the impasse of peripheral capitalism clinging to expansion of the European centre. It was the same in the end for the other Arab countries. If the oil producers among them (Algeria, Libya, Iraq and the Gulf states) believed they could mobilize their financial resources to speed up their industrialization, their ruling classes could imagine only a kind of industrialization that would offer a new outlet for the exports of developed capitalism – European, but American and Japanese too. This could only strengthen the tendency towards worldwide expansion and not offer a decisive step towards an autocentric national or regional development. Once the crisis had come, this closer entanglement proved deeply catastrophic, as is evidenced by the external debt, rudely aggravated by the conjuncture of stagnation and the impact of the American counter-attack (high interest rates). In these circumstances, Saudi Arabia, Washington's traditional client, as might be expected, has opted for unconditional support to the financial and monetary system that is the instrument of worldwide expansion, and the counter-attack aimed at restoring American hegemony. If there has been any attempt at autocentric development, it has been incomplete, erratic, limited by the very character of the ruling classes of the progressive countries engaged upon it, whether they were oil producers (Algeria and Iraq) or not (Egypt and Syria). What should be noted here is that these attempts, supported by the USSR, have been fought by the West as a whole, Europe included.

To what can one ascribe this European refusal to envisage any relations with the Arabs and Africans other than neo-imperialist relations, whether they are chiefly open to American and Japanese competition (above all when the local partner insists) or relatively reserved for the Europeans?

An examination of Europe's structural and conjunctural position in international competition sheds light on this question. Europe covers the deficit on its relations with the United States and Japan by the surplus on its exchanges with the Third World and the countries of the Eastern bloc. To remain a player in the worldwide game, Europe has to maintain unequal relations within the sphere of its particular dependencies. Europe has found the main outlet for its expansion through modernization of its own peripheries (Southern Europe to be precise) and its own internal modernization. Unlike the United States and Japan who export their capital more widely (especially to Latin America and South-East Asia) in order to dominate the process of export-oriented relocation of industry in the Third World, Europe is open to massive importation of Third World manpower necessary to keep up with the rate of its internal expansion. It is also not by chance that, in the main, this immigration is precisely by those in areas of European dependence (the Arabs, the Africans, the West Indians) much more affected than Latin America and South-East Asia by the unequal capitalist development that Europe's strategy entails. It is now well known just how far this immigration has created a political atmosphere inimical to improved relations with the Third World. Finally, Europe, with a paucity of natural resources, in comparison with the United States, attaches much greater importance to securing its supplies. Having renounced autonomous military powers, Europe has condemned itself to dependence upon American good will, and relies only on its rapid intervention forces (directed against the Third World of course) that are now almost entirely the essence of the European military vision.

All this inspires little confidence in the European talk of the Third World, along the lines of a Euro-Arab–African 'trilogue' . . . We should not necessarily go so far as to conclude that it would be better to be dependent directly on the masters of the world – hegemonic imperialism – than on its lieutenants. That would leave out of account the military dimension of the problem, and rule out the possibility of internal change, perhaps less difficult to imagine in Europe than in the United States.

The intensification of Euro-Arab relations that occurred in fact after the Second World War must be reassessed in this context. It is not even necessary to draw a detailed picture of these economic relations as they are today, or as they have developed in recent history. It is enough to reiterate, as is well-known, that these relations are highly intensive in all fields. In the field of commercial exchanges, the flow from South to the North, namely Europe, assures the North the major part of its energy supplies. The flow from North to South is also significant for the Arab region: Europe is second in meeting the Arab agriculture and food shortfall and first in meeting the import requirements in producer goods for the Arab countries. This means that the relations are not only important quantitatively (revealing growth rates after the Second World War faster than the overall rate of growth in world trade) but also qualitatively crucial for both sides. The commercial exchanges are reinforced and completed by financial flows, especially since 1973 when through the recycling of part of the surplus of some Arab countries (but less and less elsewhere) some of the surplus has been invested through Euro-Arab financial institutions. These flows have considerably speeded up the transfer of technology, or to be more precise, the sale of turnkey factories. The earlier contribution of the Arab world to the creation of the labour force in Europe was significant; it has now become of vital importance. This migratory flow from South to North, although slowing down in the current crisis, seems destined to play an increasing role in the long term.

Does the crisis open new and different prospects for Euro-Arab relations? How will the conflict of economic interests henceforth between Europe and the United States be resolved, or the East–West and North–South conflicts? We have seen that the complexity of the issue stems from the following three factors at least. First: Europe is not (or not yet) a unified political and economic power. Second: it has no autonomy in defence and without this it is impossible to enjoy real political and economic autonomy. Third: Europe, by virtue of its history and the pattern of social and political forces that results, is fundamentally bound by an imperialist conception of its relations with the Third World and so far has envisaged no relations other than those based on the necessarily unequal pursuit of capitalist development. Moreover, this view is largely shared by the electorally dominant political currents, of the right or the left. This view could be reversed only in the event that through significant internal social transformations (apparently not on the agenda), Europe were able to embark on the path of 'alternative development' with non-capitalist prospects.

In such circumstances the conflict of economic interests between Europe and the United States would remain limited and subject to constraints common to the imperialist partners. The United States has successfully exploited Europe's weaknesses to mount a counter-attack to reestablish its hegemony. The purpose of this is to reconstitute a unified front in the offensive against the South, by giving preference to common imperialist interests. Unfortunately, there is no lack of evidence of this sense of community of Western interests. The common rejection of Third World demands for the New International Economic Order (a proposal formulated in the mid-1970s), the common front organized behind the IMF, using the debt weapon to subject the Third World to the objectives of the 'recompradorization' necessary to establish favourable conditions for a renewal of worldwide capitalist expansion are among the testimonies of Europe's 'rallying round' the American theses. In this context, the conflict of varying Western economic and financial interests (the dollar and the international monetary system, the opening up of Eastern and Western European trade, and so on . . .) is kept within the common Western front against the Third World. It is just as obvious at the political and military level: the establishment of Western rapid intervention forces (against the South), the shift of NATO forces to Spain and Italy, more directed against the Arab world than against a hypothetical Soviet attack, are clear signs of this common political front. Twice in the autumn of 1985 and at intervals of a few days, American bases in Sicily were used against the PLO in Tunis and then in Egypt. Virtually unconditional support for Zionism is part of the same scheme.

Post-war expansion was always characterized by the deployment of a plan for national bourgeois development throughout the Third World, and especially in the Arab region.

We shall not repeat our analysis – available elsewhere[8] – of this project that we have described as an attempt to build a modern, industrialized national economy, driven and controlled by a national bourgeois state. This plan, fundamentally capitalist – and thus in no way 'socialist' – envisaged no 'delinking' but on the contrary a tighter and more effective integration into worldwide expansion. In other words, it was a plan for 'development' in the sense of the conventional ideology of development prevailing in the 1950s, 1960s and 1970s: of hoisting oneself from the (so-called 'underdeveloped') status of periphery into that of a new centre, crystallized as such and as an equal partner in the world system.

Thus, in the Arab world, from 1945 to 1970, along with the rise of the national movement, there has been an apparent attempt at crystallization of an Arab national bourgeoisie, or Arab national bourgeoisies, believing itself capable of forming a hegemonic political and social force at national level and becoming an equal partner in the world system. When one says 'equal' this does not betoken equality on all levels. Nobody would say that Italy and the United States are 'equals'; but they are equal partners in a certain sense, namely that there is no fundamental barrier to the less advanced partner's eventual catching up with the other.

This attempt at bourgeois crystallization would entail control over the labour force, which entails balanced development of agriculture and industry so that local agriculture is able to supply the essential elements for reproduction of the labour force. It would also entail control over markets, at least the internal markets, and control of natural resources. Nasser's Egypt had begun this resumption of control of natural resources, even if these natural resources are much more modest than those represented by Arab oil. It would also entail control over the financial circuits and an attempt to master technology, albeit imported technology. All this seemed possible at the time; at least the discourse of the Arab national movement suggested that it was. History has proved the contrary.

This failure raises several questions. The Leninist thesis, as we understand it, argues in effect that since the end of the nineteenth century the Third World bourgeoisies were no longer capable of crystallizing into national hegemonic classes integrated into the world capitalist system. Although, in the previous history of capitalist development, there had been no contradiction in the formation of new national bourgeoisies (in Italy, in Germany, etc.), and their integration into the system of worldwide expansion, the contradiction had arisen subsequently.

Very briefly one can give some examples of the weaknesses of this construction, whether it be 'socialist', 'Ba'athist' or 'Nasserist', or attempted under the aegis and with the blessing of imperialist circles in certain newly 'industrialized countries' of Latin America and East Asia. External indebtedness and current bankruptcies also reveal the frailty of these liberal experiences. This general frailty permits imperialism to go back on the offensive; and this offensive is called 'recompradorization'. Its Arabic name is *infitah*, presented as an 'open door', as if the Arab world had ever in its history been closed in on itself. This plan for recompradorization is implemented through restructuring plans imposed by the International Monetary Fund, with the strong support of Saudi Arabia. The plan also calls for political and military intervention; this is why imperialist support for Zionist expansionism is more overt than ever.

The United States has learned the lesson of the Shah's downfall that they cannot rely on local powers in the Third Word, even if these reactionary powers are devoted to them, and that it is better to rely on agents of intervention more structurally and historically bound to imperialism. The same thing is happening in Southern Africa with South Africa, which until recently (before the summer of 1985 marking the opening of the final chapter in the struggle for liberation in this country) believed itself, like Israel, in a sufficiently strong position to compel neighbouring national powers to accept 'compradorization' and surrender. NATO's intervention in Lebanon was intended to turn the comprador Lebanese state into a second Israel, fascist, falangist, pro-Western and obviously pro-Israeli.

If the Arab national plan has proved impossible to achieve, as is demonstrated by its current degeneration occasioned by the crisis, the failure is due also to internal causes (the bourgeois character of the plan) and to the fact that the West, far from supporting the development, has fought against it and continues to do so. Here again we shall not repeat what we have written elsewhere about the plan's internal contradictions, its historical limits and extreme vulnerability, which have in the end led to its failure. But we would emphasize the point too often hidden that the 'internal causes' have not operated in isolation, or in an atmosphere conducive, or even neutral, to the plan. On the contrary, the world system – central domination, with or without hegemony (American in this case) – is far from being favourable to homogenization of the system by the gradual crystallization of new partner centres (as all versions of the 'stages' of development theory suppose), but has had rather the reverse effect of further reproduction of the centres/peripheries asymmetry. This point is of major theoretical and practical significance for us.[9]

In the Arab region, the plan is based on an attempt at unitary national construction, of which Nasserism was the highest point. Again we shall not repeat the thesis we have outlined elsewhere on the Arab nation and Arab nationalism. The distant past is of great significance here despite the eight or nine centuries of degeneration that followed – a past including the character of social formations in the Arab world in its first glory (the first three or four centuries of Islam), marked by the unification of the dominant class on the basis of statist/mercantile centralization of the surplus (in contrast with European feudal fragmentation), and hence the unification of culture and language. The renaissance that came on the horizon from the beginning of the nineteenth century was built progressively on Arab unitary nationalism, breaking with Ottoman influence, and Pan-Islamism. But the arrival of the necessary elements for the plan's implementation, namely liquidation of the Ottoman Empire and British and French colonizations, set up obstacles to the plan. The Arab states, one by one, regained their independence but in disunion. Gradual reinforcement of these new realities, far from narrowing the differences that had been opened in the preceding centuries and worsened by colonization, served rather to entrench the differences. The Arab bourgeoisie began to be aware of its possible collective emergence only when it had given way to a series of local bourgeoisies, each integrated into the world system.[10]

For all kinds of reasons, some general and fundamental (the West's hostility to the emergence of new centres in Asia and Africa), others more specific to the region (the markedly popular dimension of the national liberation struggles, conflict between states, the Palestinian question, to which we shall return), the hostility of the capitalist West was unyielding and particularly violent. Is it necessary to recall the facts: the 1956 aggression against Egypt, the decision taken by the Americans in 1965 to go to war to bring Nasser down, and put into effect in 1967 by Israel and its sleeping partners, the prolonged Algerian war (1954–62), the invasion of Lebanon in 1982, the annexation of the Golan Heights and the West Bank of the Jordan by Israel, and so on.

What is important to note here is that while this constant conflict between Arab nationalism (bourgeois though it be) and imperialism has been one of the ways in which the USSR escaped the isolation to which the Atlantic alliance sought to confine it, Europe has never dissociated itself from the United States in the conflict. The supply of Soviet weapons to Egypt in 1955 clearly marks the Soviet Union's entry on the Arab scene. The parameters of this intervention, necessarily an aspect of the Soviet Union's global strategy (whether of cold war, or of detente), and the associated ambiguities (particularly as regards the 'transition to socialism') are important topics of discussion that we shall not go into here.[11] On another tack, Europe, one need hardly recall, after dragging out its efforts to hold on to colonial possessions (the Algerian war and the Anglo-French attempt against Egypt in 1956 are evidence of this) simply walked off-stage to leave the American policeman and its Israeli subordinate to act for the entire West. At least, until 1973, when the 'oil crisis' sharply woke up the Europeans and reminded them of their own vulnerability and of the selfishness of the United States. But what has Europe done since? Its 'comeback' in the Orient coincided with the decline of the Arab nationalist plan; Europe was happy to show a good face – for the sake of business – to the new ruling Arab forces, especially the most reactionary and the most susceptible to accepting the compradorization under way. This cannot be said to be 'supporting an Arab attempt at autonomous development', but merely as enrolling as a partner – albeit a trading competitor – in the American plan for the region.

The post-war upsurge followed by the crisis, far from narrowing the North–South gap, has widened it, setting the two shores of the Mediterranean further apart than ever, through the closer integration of Southern Europe on the one hand and the rejection and sinking of the Arab plan on the other.

How do these conflicts relate to the East–West conflict? Our thesis, expounded elsewhere, is that the East–West conflict is artificially kept alive by the United States precisely in order to rally Europe through the Atlantic alliance into a common offensive against the South. In this sense the conflict is largely exploited as a means of blackmail. This does not prevent the East–West conflict having its own autonomy (the conflict of the two superpowers and the means they deploy to dominate their respective spheres of influence, the arms race and the American strategy of wearing out the USSR's strength by this means, etc). Acknowledging these dimensions in no way means accepting ideological discourse on the capitalism/socialism conflict.[12]

How does one situate the East–West conflict in the context of the broad offensive of imperialism against the South in general and the Arab world in particular? It is not our purpose to offer an analysis of the character of the two social systems, nor shall we discuss the general issue of war and peace and its relationship to the character of these systems. It might, in a word, be too provocative: capitalism does not harbour war as the cloud does a storm; nor do we believe that the Soviet system necessarily harbours expansionist desires. Both systems harbour war and peace within them. We are closer to Clausewitz's view that war is 'nothing but the continuation of politics by other means'. War is the means of overcoming certain contradictions and as such is not inherent in any one social or political system.

One must consider the East–West conflict not in ideological terms, as is so often done in the West, but in pragmatic terms. It is a matter of seeing who, in the current conjuncture, takes an aggressive stance towards the other, and why? The thesis put forward here is that the East–West conflict was artificially kept alive by the American administration from the closing years of Carter's term – in an action that was pursued even more blatantly by President Reagan – with the aim of compelling Europe to rally to the Atlantic spirit, for the purpose of restoring the American hegemony that had been challenged by the economic development of Europe in the last 30 years. We might add that this rallying is aimed at collective confrontation with the South. There are various indications of this. In the West, the media often portray this as an East–West conflict, in which the Soviet Union is currently embroiled through Syria, and in the past through Egypt. This in no way corresponds to the truth. A further indicator is: the shift of NATO's military strategies towards the South and the placing of missile bases in Sicily not aimed at the Soviet Union but at the Arab world. So under cover of an intensification of conflict with the East, the conditions are created for aggression against the South. The Mediterranean is no longer NATO's southern flank against the Soviet Union, but NATO's central flank against the South. The strategy seeks, therefore, to recompradorize the enormous space that covers, among others, all the Arab and all the African peoples.

Seen in this perspective, Euro-Arab relations are unlikely to develop in a way favourable to the liberation and progress of the Arab peoples. Euro-Arab relations are currently at a conjuncture that is highly unfavourable to Arab popular interests. On one side we have a Europe that after erratic changes of heart towards the Arabs between 1973 and 1980 has totally gone over to the American and Zionist plan for the region. And on the other side there is the Arab world of *infitah*, a disintegrating Arab world where a half or more of the powers are already openly compradorized.

Does this mean there is no room for any other economic and political relations between Europe and the Arab world? There is, but on condition that the relations are within the prospect of reinforcing the autonomy of the states and peoples in regard to dominant American imperialism. In such a perspective of widening autonomy of the European and Arab states and peoples, one might imagine that some kind of mutual support is not impossible, despite the past and despite the difference in levels of development. This is the prospect of a non-alignment reinforced by a European non-alignment and a restored Arab non-alignment.

The Arab nation and the Israeli challenge

The Israeli challenge is not the product of a particular exogenous force, Zionism in this instance. Zionism would never have seen the light of the day without imperialism. Even before the notion of Zionism emerged, imperialism had always planned the creation of a 'European' state on Egypt's eastern borders, capable of ensuring the West's control over Suez and of ruining any chance of an Egyptian renaissance, the precondition for Arab unity. Without the British mandate in Palestine there would have been no State of Israel.

Zionism, whatever the internal reasons in (Central and Eastern) Europe for its appearance, was immediately enlisted into the imperialist plan and has never deviated from it, as the Zionist leaders themselves have often admitted. A plan as grotesque as that of 'restoring the frontiers of two thousand years ago' would have been deemed criminally fantastic in other circumstances. Does one hear of the Arabs claiming Spain, which was Muslim even longer than, and not so long ago as, Palestine was Jewish? Does one hear of the Romans reclaiming Gaul? What is the right of hypothetical 'descendants' of a people in comparison with those of the actual occupants of the country over 20 centuries? Do we have here to accept the principle of *lebensraum*, or 'living-space' so essential to some as to justify them driving others out of their own country, a principle formerly propounded by the Nazis?

The reason for the West's interest in the Zionist plan was that it created a 'rapid intervention force' – to use the strategic terminology of our own day – permanently established close to Suez and the oil. It was, therefore, essential that Israel stubbornly refuse any 'concession', any recognition, even a partial one, of the right of Palestinians, and even of their Arab neighbours, to existence as an autonomous nation (or nations). The refusal 'to accept established frontiers', the constant raising of the stakes in expansionist claims (half of Palestine, then the whole of it, then Sinai and the Golan Heights, Southern Lebanon, the West Bank of Jordan . . .), these are the means by which Israel, in the eyes of Western imperialism, justifies its reason for being, as a perpetual pretext for military intervention for the sake of the possible needs of the West. Would peace and recognition of this country's place in the Middle East not mean the end of Zionism? Abandonment of the external support essential to its survival?

That is why Zionism, obeying its own logic, has chosen to play the card of maximum East–West tension, in such a way as to heighten its usefulness to the extremist leaders in Washington. En route, Israel, by providing a test bed for Western and Soviet armaments (the Arabs have no choice but the latter), serves an extra purpose in the eyes of the Pentagon strategists.

In these circumstances, one must not underestimate Zionism's role in the camp of opponents of autonomous development for the peoples of the Third World in general, not to mention opponents of prospects of socialism, or even more prosaically of detente and prospects of peaceful East–West coexistence. The role played by Zionism in the 'anti-Third World' campaign aimed at ruining hopes of improved relations between Europe and the Third World, and facilitating Atlantic realignment under United States aegis should also not be underestimated. But in a further ideological anomaly, Israel has, despite the 'Reaganite' option, been able to retain the active sympathies of the Second International among others by manipulation of the constant blackmail of anti-semitism! Is this yet another demonstration of how difficult the Europeans find it to escape their 'Eurocentrism' and imagine that one of their problems does not necessarily concern their victims?

We should add that Zionism has entailed a fantastic ideological retreat even within Europe. Five centuries of the Enlightenment followed by the struggle of the labour movement had in the end won acceptance for 'the right to be different'; that is to reconcile respect for Judaism and an undeniable culture and national constituency of the Jewish communities in the European nations. These communities had – if this was necessary – proved their belonging by a remarkable contribution to European progressive thought. Zionism, by postulating a 'Jewish nationality' and a 'dual allegiance' of Israeli citizens, was in fact overwriting this progress and bringing back the attitudes of the Middle Ages. It has already managed to shift the 'Israeli electorate' to the right, into the

Atlantic alliance (although such an electorate was totally unacceptable in the Europe of a few decades ago). Zionism is nurtured by the rebirth of its twin brother of anti-semitism, and uses all the means it can mobilize to this effect.

What is at stake in the crisis and in autonomy for the Mediterranean region

The current crisis reveals more clearly than ever the sharp choices that must be made. In simple terms one might say there are two families of response to the challenge. There are right-wing solutions that may also be implemented by governments supposedly of the left, who finally submit to the constraint of 'international competitiveness'. With this outlook they envisage global domination by the North over the South, and hence by the American and European Western world over the Arab world, through control of relocation of industry. This is, in fact, the policy implemented by some and the policy behind the offensive against the Arab world, and the Third World in general, with the aim of ensuring 'compradorization' of the Arab world. By contrast, a left-wing solution entails a transformation in the historical and social outlook, not only in regard to Euro-Arab relations but also at the level of internal social constructions in the European and Arab worlds. It implies that on the European side the labour movement and the left opt clearly for an alternative development, and not merely restoration of the mode of development characteristic of post-war prosperity; an alternative development putting more emphasis on expansion of non-market production. The option for this alternative development entails substantial social and political transformations in the West, changes for which the historical models of socialism, whether of the Second or Third International, do not provide an adequate answer. It is a socialism yet to be invented that would, if developed, provide an effective way of founding relations with the South in general and the Arab world in particular on an alternative basis. These relations would permit effective national and popular development in the South. Those relations that have until now been a barrier to Arab development would become a positive encouragement for it.

It is, therefore, evident that when one talks of negotiations, of Euro-Arab dialogue, one must know where one is standing. Within the outlook of giving the capitalist system a boost through these right-wing solutions of relocation of industry to the South dominated by imperialism, or within the outlook of social transformations conducive to genuine development in the South and especially the Arab world?

The question we are raising is whether a better world order is really possible. Doubtless the countries of the region, those of Southern Europe as well as the Arab countries, are neither isolated from one another, nor cut off from the rest of the world. But recognizing this fact, the reality of worldwide expansion, not only of the economy but also of technology, science, culture, ideology, politics and strategy, is in the end no more than to admit something quite conventional.

This worldwide expansion of capitalism has always been unequal and will probably remain so between the countries of central capitalism and those of subject peripheral capitalism.

For the region, in any event, recognition of this worldwide expansion is also the recognition that the 30 years after the Second World War considerably changed the profile of the national societies and consequently the forms and degree of their entanglement with the world system. Italy, Spain and Greece have undergone rapid industrial development, which has substantially altered the position of these countries in the international division of labour and brought them closer to their wealthier and more highly industrialized partners in Northern Europe. By the same token these evolutions have perhaps accentuated the breach between the north and south of the Mediterranean. To be sure, the Arab countries have also undergone tremendous changes in the 30 years after the Second World War. The attainment of independence for the countries that were still formally colonies in 1945, the rise of Arab nationalism and aspirations to unity, industrialization and development of national administrative and educational systems, have substantially changed the face of these countries. They are no longer as they were half a century ago – essentially rural countries – integrated into the international system by colonization and even settler colonialism (in the then French North Africa). Does this mean the process of modernization north and south of the Mediterranean belongs to the same family of phenomena in the world of today? And that this modernization will sooner or later bring the societies south of the Mediterranean closer to those of the north?

For many, perhaps even for the predominant forces within the region and outside, no development is possibe other than that following integration into the 'normal' trends of the international division of labour. It merely remains to adjust to this division, or submit to its constraints. Worldwide expansion is regarded as an implacable, almost supernatural, force. This is certainly the view taken by the World Bank, the International Monetary Fund, the Organization for European Economic Co-operation, the European Economic Community and the region's governments, or the prevailing tendencies in these governments, as perhaps in most of the world's governments. One must not forget that the prevailing tendency in the world, and hence in the region, is to regard development as nothing more than this fairly passive integration into the international division of labour. But then one may wonder if there is frankly any room to speak of for co-operation and possible autonomy in any region – the Mediterranean in this instance. Are there not merely international economic relations as they were long before there was any mention of co-operation? Investment of capital, the sale of goods, or turnkey factories, be it on credit or with an associated transfer of technology – does this deserve the name of 'co-operation'? Is this not merely double talk of the period after the Second World War, when the blunt old expressions of 'colonization' and 'capital investment' were replaced by sweet talk of 'co-operation', although the net result of the operations in question was no different?

But for those like us who are trying to define the conditions for genuine co-operation under a different outlook, there is perhaps another way. It is undoubtedly very difficult and some would call it utopian. Is it possible to reconcile some of the demands of worldwide expansion with another demand that remains fundamental: national construction with the popular social content essential for many countries? Is it possible to formulate strategies and policies that will allow more scope for autonomy within the world system, to the benefit of the states, nations and popular classes that make up the nations of the region?

The conjuncture of the mid-1970s seemed favourable to this latter outlook. The war of October 1973 and the emergence of the group of oil producer states as a partner able, for the first time in the history of North–South relations, to act effectively in setting a significant price (that of energy), plus the political conjuncture associated with the end of the dictatorships in Portugal, Spain and Greece, along with other indicators of the international conjuncture – beyond the region – made it appear that the bipolarity of the post-war period was vanishing and that a multipolar world would sooner or later take its place. But the modernization of Southern Europe has come about in the framework of this bipolarity, with the European construction looking to the Atlantic alliance, while the radical Arab current has constantly fallen foul of the imperatives of American hegemony to which Europe had rallied.

The subsequent years suggest less optimism about the direction of global change. The United States has returned to the offensive. It is possible now to see that if, during the 1970s, this offensive had rather played South's and East's cards against West's cards, we would see in the 1980s a logical reversal of this strategy in the counter-attack. Henceforth the United States will play for a West totally realigned and mainly against South, using perhaps the excuse of the threat from East.

This choice – of enrolling in the implacable development of the world system or trying to reconcile some of its demands with those of a more autonomous development to the benefit of nations and peoples – must be made in circumstances of acute crisis. We are agreeing here with the analysis put forward by Giovanni Arrighi[13] on the character of this moment of crisis for the expansion of capitalism, described as a moment of 'unruliness'. Arrighi means by this that the requirements of the restructuring of capital come up against various resistances: of the working classes in the industrialized countries, who are – to varying degrees and according to their sectors – victims of restructuring in the social conditions where it occurs, but resistance too, of the states and peoples of the Third World, to what we crudely call the 'recompradorization' under way. Moreover, the crisis fits in with other evolutions since it occurs at a time when there are signs of the beginning of a decline of 'two Empires', the decline of the hegemony of the two superpowers. In the previous period of expansion, American hegemony over the entire capitalist system, and especially over the Western European system, including Southern Europe, had played a decisive role. In the absence of an economic or political supra-national state, this hegemony had facilitated the great wave of modernization in the opening-up to the outside.

What is at stake in the conflict is not capitalism or socialism in the abstract, as systems of social organization. The real alternative is worldwide development, which in the current phase supposes a return to submission to United States hegemony over the entire Western system, or a rejection of this approach in favour of a strategy of maximum autonomy. Accepting the worldwide expansion means accepting a dual recrystallization around the two superpowers, in an atmosphere of cold war spiced with hot wars (and there are such in the region) with all that cold war and hot wars can bring in a slide towards more massive destruction. The alternative is therefore: to accept worldwide development with all that this entails, or to try to implement policies of national and popular autocentric development, which will be appropriate forces in reshaping the national formations and the world system. This alternative is between the worldwide expansion or the broadening of the scope for autonomy for peoples, states and nations, that is for the benefit of the popular masses and working classes. To submit or, to the greatest possible extent, delink the fate and destinies of peoples, states and nations from the implacable demands of worldwide expansion.

Undoubtedly this debate raises fundamental theoretical questions that we can only glance at here.

For the societies of the South, is genuine national liberation with a popular bent, which is far from being achieved, still possible? Certainly the evolutions one sees taking place through the Arab region, described under the now well-known term of *infitah*, are an expression of this compradorization that is the very antithesis of development within a perspective of national and popular liberation.

For the societies of Southern Europe and Europe in general, is it possible to make a start at least on a development other than what would follow the simple logic of capital operating in the framework of the European Common Market and the world system, a development that would offer scope for social, cultural and political reforms?

Surely broadening the space for autonomy for the South and the West would also be the best way to encourage reforms in Eastern Europe? For this East is neither far away from the region nor unfamiliar with the problems posed in the region; Eastern Europe is very close, it almost has a toe in the Mediterranean. Furthermore the East is, in its way, a partner in the world system, and thereby a party to the system's global crisis, although it has its own particular crisis too. Giving scope to the East would perhaps be the best way of avoiding the dangers of a further flight, with all the political, military and ideological effects that could have on the region as a whole.

Is is possible to bring a convergence of these forces, that differ in historical origin and in the problems with which they are confronted, to the north and south of the Mediterranean and beyond, with the prospect of carving out scope for autonomy for the peoples and countries of the region?

Otherwise we are bound (are we not so already?) to a dual crystallization around the two superpowers, which are not specifically Mediterranean but have a Mediterranean view by virtue of their world vocation. The United States offensive in this direction is not only an economic offensive, whose implementation can be seen in the policies practised with European backing by the International Monetary Fund and the World Bank. It is also an offensive with a military dimension that cannot be erased from the evolutions in the region. We are not, therefore, on a path that encourages optimism about the margin of autonomy for the peoples and nations of the region. In the West, even the Mediterranean West, the short term pressures of the crisis are encouraging the resort to austerity policies that lay stress on the demands of international competitiveness, that is, submission to the familiar constraint of worldwide expansion. This does not augur well for relations between the countries in question and those of the South, in regard to the view taken of the division of labour between the former and the latter. There are exceptions, of course, there are social forces, political and ideological currents, scientific reflections, that suggest a very different alternative for the countries of the region. And these forces happily are not always without social and national resonance in the region. But it must also be noted that on the whole the crisis has emphasized the trend towards the Atlantic alliance, to support for Zionist expansionism and a generally negative attitude to the Third World. Moreover, during the previous period of relatively easy development and modernization, significant fractions of the progressive social forces of the Arab world believed it was possible to establish a modernized and industrialized national economy at the same time as increasing its integration in the world system. These fractions still believed in the possibility of becoming effective, albeit less developed, partners in the world system.

Today the failure of this plan has made the Arab world and its political and ideological forces turn back to the point that the horizon seems closed for the immediate future. The demoralization accompanying recompradorization, the devastating effects of oil wealth (and the regressive economic integration of the region in the world system, based on oil revenues for some and massive migration for others), the easy victories of the Zionist offensive supported by the West, all these have brought a renunciation – doubtless temporary on a historical scale – of the unitary and popular national aim in favour of a withdrawal into religion. But the danger of this unfavourable conjucture must not be underestimated, as we have written elsewhere.[14] If it persisted, it might enable the United States and Israel to impose their 'final solution' on the region: extermination of the Palestinians and overall subjection of the Arab regimes, a 'definitive' Mashreq–Maghreb split, by the coupling of the latter to the European train and Turkish style 'lumpen Europeanization', while the Mashreq would be delivered to the limitless break-up into powerless parastatal communities (the 'Lebanonization' of the Middle East), a means by which the West could ensure itself control over the sources of energy, seemingly the only vitally interesting 'resource' in the region.

If we opt for an outlook that rejects wholesale submission to the demands of worldwide expansion, for one that seeks to broaden the scope for autonomy for the region's nations and the people, we appreciate that we are swimming against the tide. This brave option entails the refusal of blind and boundless surrender to the supposed demands of international competition; the refusal of the crisis strategy of redeployment of capital; the refusal of the policy of collective aggression of the North against the South; of ideological and military polarization; and of the blackmail aimed at subjecting North–South relations to an artificial intensification of the East–West confrontation.

Only when – through action at national levels, of course – we have succeeded in establishing favourable conditions for the broadening of the scope for autonomy, can we usefully consider genuine co-operation between the states of the region, a co-operation conducive to strengthening their national autonomy. This means that the thesis of delinking from the world system is inescapable. This is where the fundamental questions arise, at least for the Arab countries, and they go far beyond the economic dimension: the choice lies between freezing the national autonomy of the Arab states as they are today, or the arduous construction of a unified Arab nation.

The choice also implies that one must not fail to give meaning to a term that remains vague: political and military 'non-alignment'. This is an expression that might well have become obsolete by now. For if in the past we have had at least two promoters of non-alignment among the countries of the region – Yugoslavia and Nasser's Egypt – who nowadays believes in it? It is true that in an extraordinary way one country in the region is convinced of the need for non-alignment: the Greece of Pasok. It is certainly a ray of hope. But for Europe one must admit that non-alignment entails a courageous choice: a break with the Atlantic alliance, withdrawal from NATO in fact, with the prospect of peaceful settlement of relations between the two Europes, Eastern and Western, and an immediate firm refusal to allow NATO bases to be used as rear support for American–Israeli aggression against the Arab peoples. It also entails, as we have said, a break with the kind of development regarded so far as the only possible, since this development necessarily demands inequality in North–South relations.[15]

It is , therefore, only by swimming against the tide by rejecting international competition, the Atlantic alliance, redeployment, polarization, militarization and all the related blackmail, that we can lay the foundations for a co-operation that is genuinely helpful to the peoples and nations of the region.

Within the hypothesis of genuine developments of this kind to the north and south of the Mediterranean a new chapter would open in the long transition 'beyond capitalism'. If we do not call this transition 'socialist' it is because we believe that here, as elsewhere, it could only be a conflicting compromise between three tendencies – socialist, capitalist and statist – combined contradictorily according to the modalities determined by the development of the struggles.[16]

Notes

1. Our thesis of unequal development in the transition of tributary formations to capitalism is presented in *Class and Nation, Historically and in the Current Crisis*, New York, Monthly Review Press, and London, Heinemann, 1980.

2. On the relation between capitalism and Europeanism, see 'The Afro-Asiatic vocation of Marxism', Chapter 5 of *La déconnexion*, Paris, La Découverte, 1986 (to be published in English by Zed Books, London).

3. Our thesis on Arab social formations in history and the question of the Arab nation is developed in *The Arab Nation*, London, Zed Press, 1978.

4. Our viewpoints on the long cycles of capitalism, post-war expansion and the contemporary crisis have been presented in our contribution to Amin S., Arrighi G., Frank A. G. and Wallerstein I., *Dynamics of Global Crisis*, New York, Monthly Review Press; London, Macmillan, 1982. For the conjuncture of years of crisis and its evolution, see also our bibliography in that work; and articles 'Développement autocentré, autonomie collective et O.E.I.N.', *Africa Development* No. 1, 1978; 'Du Rapport Pearson (1970) au Rapport Brandt (1980), ou la crise de l'idéologie du développement', *Africa Development* Vol V, No. 3, 1980, Dakar; 'Il y a trente ans, Bandung', Cairo, April 1958, mimeo; 'Les conditions d'une Sortie à gauche de la Crise', 1983; 'La Crise, le Tiers Monde et les Relations Nord–Sud et Est–Ouest', 1983; published texts also included in *'L'Echange Inégal et la Loi de la Valeur*, new edition, Paris, Anthropos, 1986.

5. Giovanni Arrighi's thesis on Southern Europe is published in this work.

6. See our studies 'L'Association eurafricaine', *Eurafrica*, April–May, 1975; 'Les perspectives internationales de la localisation industrielle, un point de vue arabo-africain', *Options Mediterraneennes*, No. 36, 1976. Also various papers on Euro-African relations collected in the new edition of our work *Sous-développement et Impérialisme en Afrique*, Paris, Anthropos, 1986.

7. On relations between Europe and the Maghreb, see: *La Coopération CEE–Maghreb*, Tunis, 1979; *L'Avenir du Maghreb dépend-il de la CEE?*, Rabat, 1981; Dialogue sur la Méditerranée (in Arabic), *Shuun Arabia*, No. 21, 1982. The papers indicated in this and the previous note are complemented by some studies on Afro-Arab co-operation ('Le Pétrole et les Relations Afro-Arabes', 1974; 'Brèves notes sur la Coopération Afro-Arabe', 1979) collected in *Sous-développement et Impérialisme en Afrique*.

8. On the plans for the national bourgeois state in the Third World and their crisis, see the article cited above (Il y'a trente ans, Bandung) and *La déconnexion*.

9. Our thesis on the subjection of Third World bourgeoisies to compradorization and the end of the era of national bourgeois attempts at the periphery of the system is explained in *La déconnexion*. It also appears in some texts written in Arabic (especially the work, *The Crisis of the Arab Society*, Cairo, 1985).

10. See *The Arab Nation*.

11. For a critique of the theory of the 'non-capitalist path' in the Third World, turn to *La déconnexion* (and *La Crise du Moyen Orient dans sa Perspective Mondiale*, Louvain, 1983).

12. Our thesis on the question of socialism and transition 'beyond capitalism', especially the thesis of the conflict of three tendencies (socialist, capitalist and statist) in post-capitalist societies, is presented in *La déconnexion*.

13. Arrighi's thesis is developed in his contribution to *Dynamics of Global Crisis*.

14. A viewpoint presented in our work in Arabic, *The Crisis of the Arab Society*, Cairo, 1985.

15. An excellent plea in favour of European non-alignment is made in CEDETIM, *Le Non-Alignement*, Paris, La Découverte, 1985.

16. We return here to our thesis of three tendencies in post-capitalism (see note 12 above).

2. The Future of Southern Europe: Canada or Puerto Rico?

Southern Europe undeniably constitutes a distinct region within Europe and the developed West in general. Portugal, Spain, Greece and Italy share a number of common features which sharply differentiate them from their neighbours to the north and give them a close kinship. Leaving aside the climate, the landscape or their traditional 'agricultural vocation', no one would deny that these countries have their specific 'Mediterranean' features within the European whole. But it is more difficult to grasp in what their originality actually lies. There is the late development of capitalism in these countries in the contemporary period, and, above all, the particular forms it has assumed, which were such that until quite recently they were classified among the underdeveloped countries of the world. There is also a certain weakness of civil society, based as it is on persistent regional imbalances and the survival of traditional family and other solidarity networks. Again, there is the recent past of fascist or conservative dictatorship and, paradoxically, the relatively high profile of the left in the current political system. Finally, there is an undeniable cultural originality, the most visible expression of which is the still very strong impact of religion.

Each of these various aspects – economic, political, social and cultural – affects the others in a multitude of ways. This makes it impossible to attribute the generally recognized specificity of these countries to the effect of a single factor. In many ways, such countries as Yugoslavia, Turkey and Albania belong to this group of nations, and this is not only a matter of geography. Their level of economic development, the impact of regional imbalances, and their social characteristics make them very similar to Portugal and Spain, even Italy, not to mention Greece. And while the nature of the social system clearly distinguishes Yugoslavia and Albania, and Islamic culture puts Turkey in a separate category, these countries share with the other nations of southern Europe, as indeed with the Arab world, an ancient heritage of Mediterranean civilization going back to earliest antiquity.

But although Yugoslavia, Albania and Turkey belong (and not only geographically) to southern Europe, they cannot be included in the same category as Spain, Italy or Greece because they have not participated in the development of world capitalism in modern times. Turkey has remained outside the dialectic of internal change and external expansion that transformed Europe from the end of the 15th century, because it was, and still is, outside the European cultural area. Yugoslavia, particularly its central and southern regions, long dominated by the Ottoman empire, and Albania too, largely remained outside the process of capitalist development in the West. Furthermore, for the same reasons Greece represents an intermediate case between these two groups of countries.

Portugal, Spain and Italy, on the contrary, played a leading role in the world development of capitalism. The cities of northern Italy saw the formation of the earliest centres of financial, commercial and industrial expansion. Portugal and Spain were the first European countries to embark on the colonial adventure and, through exploiting the land and its resources, and the inhabitants of countries overseas, accumulated a vast fortune, of which their northern neighbours were to be the main beneficiaries. It is not necessary to cite the immense impact on the progress of European capitalism of the intellectual, artistic and scientific revival contemporary with the Renaissance; or the role of Muslim Spain in the diffusion to medieval Europe of the ideas, sciences and techniques of Arab–Islamic civilization; or, finally, the heritage of ancient Greece and Rome or of the Christianity of the Roman state in the genesis of modern European culture. But, in modern times, Portugal, Spain and Italy were for long periods the centres of historical initiative and material and intellectual creativity in the context of the development of capitalism as a world system. And even in decline these nations still remained close in every respect to the new sources of expansion. After the 17th and particularly the 18th century, they allowed themselves to be overtaken by the countries of north-western Europe and later by the USA. But, except for Greece during the period of Ottoman occupation, they continued to participate fully in economic and political exchanges, and especially in cultural and human exchanges, within the West. That is why, while today Portugal, Spain, Italy and Greece (although this last is in a category of its own) still remain behind the most advanced capitalist countries, they cannot be put in the same category as Turkey, Yugoslavia and Albania. Even less can they be lumped together with Third World countries, as peripheral societies of the world capitalist system. Unlike the countries of Africa, Asia and South America, they never experienced colonial or quasi-colonial occupation and its destructive effects. They were provinces, which may well at certain periods have suffered from relations of domination and exploitation, but they were not colonies or neo-colonies.

Southern Europe since World War II

Since the end of the Second World War, the countries of southern Europe, have, like the others, been forced to open up to the international movements of goods, capital and technology. In the previous period, marked by the clash between autonomous national imperialisms and considerable instability in the relations of force at the international level, they had, like Portugal and Spain, been able to withdraw into themselves or, like Italy, attempt to drive back the established powers by force of arms. The effects of the war led to the political, military and economic unification of the countries of the West under the aegis of the United States, making autarky in the framework of capitalism henceforward impossible. Faced with the emergence of the Soviet Union as a superpower, the traditional rivalries between (small and large) competing imperialisms came to be invested exclusively in economic competition. At the same time, the expansion of international exchanges and investment increasingly led to the establishment of an international division of labour which came to impose itself on all nations and disrupted the coherence of national economies where such coherence existed. The countries of southern Europe thus found themselves caught up in a great movement reordering and developing the capitalist world economy in which, when all is said and done, they participated only passively.

It is striking in fact that for countries like Portugal, Spain or Turkey, capitalism experienced a new dynamism during the 1950s and 1960s, quite independent of any major change in the political and social system that had been there since the inter-war period. In other words, the revived vigour that the development of capitalism experienced in southern Europe after the end of the war resulted far more from external pressures than from domestic factors. For most countries in the region, geopolitics accentuated the objective effects of changes in the world system. Yugoslavia's difficulties with the Soviet bloc, Turkey's strategic position, the communist guerilla movement in Greece, and Spain's neutrality during the war made these countries prime targets of US policy in the region. And well before the spontaneous play of the free movement of goods and capital internationally could bring about changes in their economies, these countries suffered the impact of flows of economic and military assistance from the United States.

Compared with northern Europe and the United States, the countries of southern Europe have experienced relatively high economic growth over the last two decades, 6–7% during the 1960s and 4.5–6% during the 1970s. Growth rates were higher as one moved from Portugal to Turkey. The 1970s marked a slowing-down in expansion in almost all the countries, but still far from the virtual stagnation that has affected most advanced countries since 1975, which does not mean that the crisis spared southern Europe.

Looked at in terms of its rhythm, the development of capitalism was greater in this region than in the more advanced areas of the world. Indisputably, in terms of global product and per capita income, the gap between the countries of southern Europe and those of western Europe has narrowed. It is sometimes said jokingly that in the reign of Elizabeth II, England risks being caught up by Spain, which it overtook in the time of Elizabeth I! But this capitalist development has at least two features that distinguish it from the earlier experiences of the more advanced. First, it is occurring along with the conservation of 'archaic' structures of production and labour, especially in agriculture but also in industry, in both the countryside and the towns. Second, it is occurring in the context of a growing integration of these countries' economies into the world capitalist system, and particularly its European pole.

The proportion of the labour force employed in agriculture has fallen sharply since the 1950s, but it remains high. It fell from 40% to 18% in Italy, from 50% to 20% in Spain, from 50% to 30% in Portugal, from 57% to 25% in Greece, from 85% to 60% in Turkey and from 67% to 37% in Yugoslavia.

In the region as a whole, agriculture has undergone significant capitalist development, as evidenced by advances in mechanization and the use of chemicals and the increase in yields and the productivity of agricultural labour. In Italy, in the mid-1970s, phosphate fertilizer consumption was 40 kg per hectare (as against 100 in France), and on average there was one tractor for 17 hectares of land (as against 14 in France). And while labour productivity is lower than in France, irrigation is much more developed and was considerably extended in the 1960s and 1970s: in 1974 almost 30% of cultivated land was irrigated, whereas the corresponding proportion in France was 3%. In Spain, the number of tractors has increased tenfold over the last two decades and the quantities of chemical fertilizers used have trebled. Technological changes have, it is true, been on a lesser scale in the other countries of the region. In 1980, for every 100 farmers, there were about 17 tractors in Italy and 8.5 in Spain, but only five in Yugoslavia, two in Greece, 2.7 in Portugal and 1.7 in Turkey.

Agricultural investment has remained very low in Portugal over the recent period, and state policy has been simply to hold the country's food deficit within the limits set by global import capacity. In Greece and Turkey, capital accumulation has also marked time in agriculture. Agricultural investment there is generally state-financed, while agrarian capital flees to the towns to be invested in property. But, there too, agriculture has also experienced enormous technological and social changes, brought about by the extension of capitalist forms of production.

The most visible result of this has been the change in production. A trend towards clearcut specialization has gradually become apparent, aiming, on the one hand, at keeping food prices low and, on the other, developing profitable crops (fruits and vegetables) and meat production for the local market. Thus, while cereal production has stagnated in Portugal, maize imports for livestock needs rose from an average of 100,000 metric tons in the 1950s to two million metric tons in 1979, representing one-quarter of the country's agricultural and food imports. In Spain too, maize imports rose sharply, and by 1982 accounted for one-fifth of agricultural and food imports, with soya-bean imports, also intended for cattle feed, accounting for another one-fifth. The development of industrial livestock farming on the basis of imported food inputs has accentuated or engendered, depending to the case, an imbalance in external exchanges in agricultural and food products, although it did result from a reduction in imports of meat and milk products. At the same time, the structure of agricultural production shifted markedly in favour of meat and milk products. Thus, in Spain, while between 1960 and 1980 total agricultural production rose 78% in volume, that of plant crops rose by only 54% as against a growth of 124% for animal production. This specialization, which is to be found in the other countries of the region, was obtained essentially through a pricing policy whose aim to prevent the rise in the cost of food, and thus to protect the level of profits and growth in industry. The deterioration of the agricultural terms of trade, especially those of mass consumption items (cereals), channelled agricultural investment into profitable crops intended for the urban middle classes, at the same time discouraging signs of the intensification of production in the non-capitalist production systems, which were in the majority. The impact of this pricing policy might have been different if the land tenure structures and production systems had been changed. But the political power of landlords led the state to limit its intervention to the area of prices and to ignore the problems of the land tenure structure and the incomes of the great mass of the peasantry. While capitalism developed strongly in agriculture, it did not reduce all the agrarian structures and production systems to a common form.

Agriculture thus continues to be marked in southern Europe by a great heterogeneity of production systems and a high level of land concentration.

A census carried out in Spain in 1962, which focused on the unit of production rather than ownership, shows that farms larger than 100 hectares represented 2% of total agricultural farms and 56% of land. At the other extreme, farms smaller than five hectares represented 66% of the total but covered only 6% of the agricultural area! In so far as several farms might belong to a single owner, these figures underestimated the scale of the concentration of landownership.[1] At the beginning of the 1980s, according to Roux's estimates, there were some 1.5 million farms between 0 and 20 hectares in size, representing over four-fifths of the total number, but only one-fifth of the land; 240,000 between 20 and 100 hectares, representing 15% of the total and 21% of the land; and 60,000 farms over 100 hectares, representing 3% of the total but covering 60% of the land. 'The high concentration of land and smallholding continue to be the distinctive features of the agrarian structure in Spain', writes Roux, who observes further that:

> the reappearance of a considerable degree of rural unemployment in areas of large estates, a phenomenon unique in Europe, a consequence of the unjust distribution of land and also of unequal development . . . from which latifundia areas have suffered most.

Land concentration is still higher in Portugal where smallholder agriculture represents 78% of the total number of farms and occupies only 15% of the land, while 1.2% of farms cover over half the land.[2]

This land concentration is accompanied by a great variety of production systems. Not every large farm is necessarily capitalist, and in some regions, like Andalusia or northern Portugal before the agrarian reform, the traditional latifundium predominates. But capitalist agriculture, which is defined by the exclusive use of wage labour and the marketing of production, has advanced much further in the large and medium farms. Small and medium farms generally engage in commercial family agriculture, marked by the sole or principal use of family labour and production for the market. But the vast majority of farms are involved in part-time agriculture, as in the Maghreb countries. In Spain, in 1972, over half the heads of farms stated that their principal occupation was outside agriculture. And in Portugal, there was a considerable expansion of part-time agriculture during the 1970s. It represents 'the principal mechanism by which small-scale agriculture is being subordinated to the process of capital accumulation . . . at the same time as [is] the main means of resistance by rural society and the peasants to being liquidated'.[3] In 1980, for example, it was estimated that the income from the sale of agricultural products represented only 55% of the global income of farmers. In this situation,

> the combination of the classical productivist model of Western agricultures and an agrarian structure dominated by large farms and the survival of a multitude of smallholdings reinforces the inequalities in the distribution of agricultural income and makes a considerable proportion of farms economically marginal.[4]

Despite the development that it has experienced, agriculture in southern Europe remains the poor relative of growth. At the end of the 1970s, the average income of Spanish farmers was still only half that of people working in other areas.

At the same time, while urban industrial capitalist development was rapid, it was not enough to provide jobs for the quantity of manpower released by the agricultural sector. Industrial production rose by average annual rates of 10% during the 1960s and 5% during the 1970s. Turkey and Yugoslavia enjoyed even higher rates until the beginning of the 1980s. In these mineral- and energy-poor countries, agricultural production now represents only a small percentage of GNP: 25% in Turkey, 15% in Greece and Portugal, and 8% in Spain and Italy. But despite this strong industrial expansion, the labour force released in the rural areas by the development of capitalism in agriculture has been only partly absorbed by industry. Urban economic activity continues to be largely based on a sector of small artisan or family units, a large private services sector and an inflated administration.

These particular forms of capitalist development in southern Europe are illustrated by the strong expansion of the 'informal' economy, in every country in the region. 'Informal' economy refers to the activity of undeclared economic agents or the production of undeclared goods and services by officially registered economic agents.

The development of part-time agriculture has already been mentioned. In Portugal, in the coastal regions where over two-thirds of the population are now concentrated, agriculture has become mainly a source of supplementary incomes, making up for the low wages paid by other sectors and the rise in food prices. While wage-employment continues to spread, in many cases it continues to be conditioned by broader strategies of allocating family labour. Furthermore, the spread of part-time work, in the form of temporary work, half-time work and working at home has been encouraged by employers' strategies to decentralize industrial investment. The dualism of the labour market in Italy has long been known; it distinguishes between the permanent employees of large businesses and the casual and often undeclared workers in small firms. A survey conducted in the late 1970s, for example, showed that 40% of firms were not applying the national labour code and 80% of firms employing contract labour were paying wages that were half those of the big companies.[5] This phenomenon can also be seen in Greece where the proportion of wage-earners in the labour force is only 42% as against 72% in Spain and Portugal and 82% in other EEC countries. A Portugese economist who notes an increase in household incomes (as measured by declared expenditure and savings) in recent years whereas the statistics show a sharp fall in real wages, is of the opinion that the informal economy is leading to a growing divorce between trade unions and workers, or even the break-up of the working class.[6]

The development of the informal economy largely reflects the conservation or extension of archaic forms of production in the towns and countryside that are not very favourable to the expression of collective solidarities and a modern class consciousness. It masks the formation of a 'free' labour market outside regulation by the state and trade unions. Finally, it keeps workers in a situation in which they have several statuses that discourage their self-organization and autonomous expression in the social and political sphere, and promotes the survival of clientelist-type ties of personal dependence.

State, society and the international division of labour

Generally speaking, while capitalism has developed in the contemporary period to the extent that it has disrupted the old social fabric of the countries of the region, it has not succeeded in establishing new homogeneous structures for the economy and society. Of course, the situation in this respect varies greatly from one country to another, depending on levels of development and national history. In Italy, and to a lesser extent in Spain, recent economic growth has, despite everything, led to a significant reduction in the reserves of non-wage labour, and hence increased the social and political power of workers. Thus, in relation to Italy, there has been talk of a neo-corporatist model, in which, as in northern Europe, (big) capital and organized labour attempt to use the state to establish a social compromise around the growth and the distribution of national income.[7] Post-Francoism has also been interpreted as the move from a fascist-type corporatism to a new form based on a marked democratization of society,[8] but one that is, nevertheless, 'polyarchic, oligopolistic and class-based'. Here, unlike the advanced countries of northern Europe, however, the state plays a much more determining role in that the relatively weak organization of capitalists and workers gives it a greater degree of autonomy in relation to civil society. Quite clearly, this weakness of civil society is even more marked in the other countries of the region. In Greece, Portugal and Turkey, the widening of political participation occurred in a context marked by a great weakness of the industrial proletariat, and modernization has been achieved not through the development of powerful political and trade union organizations, but on the basis of the transformation of traditional networks of clientelism and, sometimes, the intervention of the army.

In the context of the current crisis, the function of the state becomes crucial in these countries, in that, in the absence of mechanisms for class compromise, it must mediate not only between labour and capital, but also directly negotiate a better place in the international division of labour. In the case of Turkey, this enhanced state responsibility took the form of direct intervention of the army and the establishment of a very authoritarian and repressive political system that, against the will of the popular classes, guaranteed the implementation of a programme of extraverted industrialization. Elsewhere, this policy is being attempted in the context of broad political alliances including the popular classes. The capitalist development of the countries of the region during the last two or three decades has been consistently associated with a growing opening to the outside world and an increasingly deeper integration into the world capitalist system. Initially stimulated by American aid, economic growth later rested largely on the extension of relations with the more advanced zones and in particular with the neighbours to the north.

Between 1960 and 1979, the degree of openness of the economy to the outside world, measured by the ratio of exports to Gross Domestic Product, rose from 11% to 15% for Spain, 9% to 17% for Greece, 3% to 5% for Turkey and 17% to 26% for Portugal, whereas in Yugoslavia it remained unchanged, at 14%. External, essentially US aid, income from tourism, principally west and north European, and remittances from workers who have emigrated to France, Germany and the Benelux countries consistently represented a very important element in keeping the balance of payments in equilibrium and shoring-up the import capacity of the countries of the region. US aid represented a very high proportion of imports into Turkey and Greece throughout the post-war period, and in Spain and Italy during the 1950s. Earnings from tourism covered and still cover a considerable proportion of the trade deficit in Spain and Italy; in Italy, remittances from migrants increased significantly during the 1960s, and in Spain, Portugal, Greece and Turkey during the 1970s. The size of these transfer incomes indicates the inability of capitalism in the countries of southern Europe to ensure economic and social reproduction on the basis of internal resources. Their amount has been reduced by the economic crisis, posing the choice between reducing investment or developing exports. The constraint on external competitiveness engendered by the crisis-induced reduction of transfer incomes was in any event reinforced by the development of foreign investment and commercial exchanges with the more advanced economies. The closer integration of the countries of the region into the world capitalist system is in fact not limited to participation in the networks of international income redistribution through aid, tourism and migration.

Foreign investment has been very important in the economies of southern Europe since the 1950s. It is true that their neighbours to the north-west received almost twice as much foreign investment per capita. But to the extent that foreign capital in southern Europe is concentrated in industrial sectors that are often strategic ones and in large firms dominating an environment of small units, its influence on the economy is much greater. Although state policies have systematically encouraged the inflow of foreign capital, notably in industry, numerous studies have shown that external investment has rarely had a positive impact on the industrial integration, employment, income or technological level of the countries concerned. Conversely, this investment has greatly contributed to the imbalance in the balance of payments. In Greece, for example, by the mid-1970s, outflows of capital occasioned by the 'repatriation' of profits and interest payments represented between three-quarters and all capital inflows.

Since the 1960s, Spain has greatly increased its share of the world market. According to some estimates, it has increased its outlets in developed countries by 50%, but it has doubled them in the Third World and increased them fourfold in the oil-producing countries (IMF). Greece and Turkey have increased their markets in the oil-producing countries to a lesser extent and much less in other underdeveloped countries. As for Portugal, the loss of its colonial empire meant that it lost the bulk of its markets in the Third World without having been able to develop new ones elsewhere. In every case, these countries underwent a sharp reorientation of their external exchanges towards the European Community. About half the exports of Greece, Portugal and Spain, for example, went to the EEC in 1980, and the proportion was the same for industrial exports. Even before the official entry of some southern European countries into the Common Market, their commercial integration into the European Community was already a reality. As was to be expected, exchange flows with the EEC are dominated by West Germany and France, which accounted respectively for one-quarter and one-fifth of them. On the other hand, the flows that have developed between the countries of southern Europe have remained relatively limited.

Trade with the EEC is generally in surplus for agricultural products and in deficit for industrial products. Studying the structure of the external trade of the countries of southern Europe, Giannitsis[9] distinguishes between what he calls Ricardo goods: agricultural and mineral, and Hecksher-Ohlin goods: labour-intensive industrial, and high technology goods. Such a distinction may be questionable, because the international trade theories advanced by Ricardo and Hecksher-Ohlin are alternative theories that are mutually exclusive rather than complementary, but it can be accepted for explanatory purposes. Giannitsis finds that, globally, between 1974 and 1980, the countries of southern Europe succeeded in increasing the proportion of high technology goods in their exports at the expense of Ricardo goods, with those of labour-intensive industrial goods remaining more or less constant. They can thus be seen as occupying a special position in the international division of labour, different from the more advanced countries, the semi-industrialized countries of Latin America, which are more specialized in Ricardo goods, and the semi-industrialized countries of south-east Asia, which are more specialized in Hecksher-Ohlin goods.

It is true that the structure of external trade has changed since 1960.

Specialization in primary exports, chiefly agricultural products, has given way to a specialization in exports of industrial products, consumer, and intermediate goods. Between 1960 and 1980, the share of industry in exports rose from 11% to 51% for Greece, 62% to 72% for Portugal and 34% to 75% for Spain. The relatively high figure for Portugal at the beginning of the period reflects an old specialization in the framework of the colonial empire. But textiles and clothing represented almost half the industrial exports of Greece and Portugal in the late 1970s. Although these two countries export not insignificant quantities of metal and equipment products, their specialization in non-durable consumer goods industries comes out clearly, and their industrial growth has simply broadened their initial comparative advantage. Spain, on the other hand, has succeeded in establishing a more diversified structure of exports: metal products, transport equipment and machinery, each represent a quarter of industrial exports. Of the three countries, the industrial exports of Greece are most concentrated on non-durable consumer goods and the range of industrial activities is the smallest. At the other extreme, Spain is the country whose industrial exports are the most diversified towards the upper end of the range and whose range of activities is the most open. From this point of view, Yugoslavia is closer to Spain while Portugal, and above all Turkey, are more like Greece. This difference in structures of production and exports is of course reflected in a difference of roles, or at least of places, in the international division of labour.

In fact, it is Spain (and of course, Italy) which has greatly increased the proportion of high technology goods in its exports, which moreover are still focused on Hecksher-Ohlin goods. The other countries continue to be strongly specialized in these same goods. But the fact remains, that in southern Europe the structure of industrial production continues to be dominated by consumer and intermediate goods industries. In the late 1970s, as a percentage of crude output, capital goods were only 10% in Greece, 16% in Spain and 13% in Portugal.[10] The foodstuff and textile industries are the main activities in Greece (24% and 15%) and Portugal (23% and 13%). In Spain, the foodstuff industry is also important (16%), but steel (13%) and transport equipment (16%) have become leading activities.

Southern Europe and the EEC

The entry of Greece, Spain and Portugal into the EEC is subjecting their industrial apparatus to direct competition both from the more advanced economies of the region and from the semi-industrialized Third World countries. Integration into the Common Market, especially for such countries as Greece and Portugal, not only makes it more difficult to maintain traditional specialities but also creates obstacles to the search for strong positions in new industrial activities. Comparative advantage can no longer be based on the price of labour; the semi-industrialized countries of the Third World are now more competitive for labour-intensive products. Despite its limits, economic growth has led to the reduction of the structural surplus of manpower, and political democratization has removed the repressive regulation of professional relations. At the same time, the search for new specialities in more sophisticated activities has to be carried out in the framework of economic and technological free trade at the level of western Europe, which makes the implementation of voluntaristic policies extremely hazardous. Southern European countries' adoption of the common external tariff, in most cases lower than their own tariffs, and the freeing of trade with the EEC countries will quite probably end in an increase in net imports of industrial products and a worsening of their trade balance deficit. The result will be increased pressure not only on the sectors of traditional specialization, such as textiles, clothing and shoes, but also on intermediate technological activities, such as car manufacture, steel and shipbuilding, that are particularly important in Spain, but are also present in Portugal and Greece. Despite the envisaged transitional measures, membership of the EEC will thus make it vital to implement radical policies of industrial restructuring. In a context where equipment goods and technologically sophisticated goods can be freely imported, notably from the more advanced EEC countries, it is difficult to see how such restructuring policies could develop new, more dynamic branches of activity, especially for countries whose economic size is, after all, modest. Restructuring would then mean a modernization of existing sectors with a view to improving the level of productivity, and would necessarily involve negative effects on employment. Furthermore, integration into the EEC threatens to increasingly penalize small and medium industrial firms, which will have much more difficulty in facing up to international competition.

Finally, the experience of 'community development' over the last two decades augurs rather badly for the prospects of reducing regional imbalances, both in the countries of southern Europe and between these and their more advanced neighbours to the north.

Thirty years after the signing of the Treaty of Rome, the Common Market is a long way from having reduced the gaps in development between regions. In Italy, per capita income in the Mezzogiorno is still only 60% of the national average, which is more or less the same as it was in the late 1950s. This result, true to the logic of the uneven development of capitalism, can also be explained by the absence of a real community policy of regional balance. The resources of the EEC institutions responsible for regional intervention have benefited the southern regions of France and Italy to a lesser extent than they have Ireland and the old industrial regions facing difficulties in the Community.[10] And not only has the EEC's agricultural policy paid no attention to agrarian structures but, above all, has kept up the prices of non-Mediterranean products (cereals, sugar, milk, meat, butter).[11] With the entry of Spain, Portugal and Greece, however, gaps between regions are becoming much greater within the EEC, of the order of 1:15 between the poorest regions of Portugal and the Hamburg region, for example. It is clear that in these conditions, the already considerable regional imbalances internal to the three new member countries are likely to persist or even worsen. In Greece, one-third of the population and one-half of

Gross Domestic Product (GDP) are concentrated on 3% of the national territory; and in Portugal, nine-tenths of GDP and four-fifths of the population are concentrated in the three coastal regions in the north; while the ancient dualism in Spain between Catalonia and the Basque country and the rest of the country is well known.

These likely consequences of integration into the EEC are clearly aggravated by the current crisis of the capitalist world economy. In the countries of southern Europe the reduction of external investment or its diversion towards low-wage areas, the halt to emigration, the fall-off in tourism and the contraction of external markets are exacerbating the slowing down of growth, and the inflation, unemployment and external indebtedness that arise from the contradictions of local capitalist development. Similarly, the impact of the restructuring called for by entry into the EEC is augmented by the deflationary policies implemented by left-wing governments in order to try and get out of the crisis. It may seem that in the medium term, the combination of all these factors will have the consequence of accentuating the perverse forms of the development of capitalism in those countries. The diversity of structures of production in all sectors of activity and the informal economy will, in this context, not only have to survive but to spread in order to enable the local economy and society to absorb the shock waves precipitated by the abrupt exposure to external competition.

But the integration of Spain, Portugal and Greece into the EEC is not a choice dictated from outside and in Italy no one questions membership of the Community. And the more or less formulated candidacies of Turkey, and even other non-European countries (the latest being that of Morocco) are evidence of the attractive power of the European Community in the Europe–Mediterranean region. It has rightly been said that for such countries as Portugal and Spain, integration into the EEC ought above all, beyond strict economic considerations, to make the break with dictatorship irreversible along with the autarky, relative though it may have been, which undermined it. Even from the economic and social point of view, however, the attractive power of the EEC cannot be attributed to the blindness of the public opinions and even less of the governing elites of southern Europe. Both may have illusions about the possibilities opened up to them in the long term by an integrated Europe in, for example, the area of the free movement of labour. But it is very likely that in Portugal, as in Spain and Greece, most sections of society are clearly aware of the place their country will be able to claim in an enlarged Europe, and accept its limits and constraints. The fact is, the alternative choice that may be conceived in the framework of their ideological horizon, is neither that of a non-capitalist development within national borders, nor of a capitalist development that turned its back on northern Europe and was based on integration among the countries of the Mediterranean basin. The only alternative choice to integration into the EEC then remains the continuation of the development model of the last two decades, a model marked by a strong dependence on the United States and major regional, social and sectoral imbalances. Compared to such a model, membership of Europe makes it possible, first, to moderate dependence on the United States and, second and above all, to be part, even be it as a poor relative, of a metropole of the world economy. Between dependence alone and dependence as part of a dynamic and powerful grouping, the choice has certainly been made easier by the desire to strengthen the newly acquired democracy and by the cultural proximity to northern Europe. But it has perhaps also been guided by the hope of a greater prosperity. How far increased prosperity may result from an improvement in productive performances in the framework of the Common Market or from a simple policy of social transfers within the EEC remains an open question: Canada or Puerto Rico?

Notes

(All the references quoted refer to papers presented to the colloquium organized by UNITAR and the IASM at Naples in October 1983 on 'alternative development strategies in the Mediterranean basin'.)

1. Bernard Roux, 'Agriculture et internationalisation du capital en Espagne'.
2. Afonso de Barros and Armando Trigo de Abreu, 'Agriculture et crise économique au Portugal'.
3. Ibid.
4. B. Roux, op. cit.
5. M. Ikonikoff, 'Economie souterraine et système économique mondial, l'expérience des pays de l'Europe méditerranéenne'.
6. M. V. Cabral, 'L'économie souterraine au Portugal'.
7. G. Arrighi, 'From rivalry to American hegemony: implications for the political economy of the southern European countries'.
8. S. Giner, 'From corporatism to corporatism: the political transition in Spain'.
9. T. Giannitsis, 'International specialization of the European newly industrializing countries: evolution and differentiation in the period of crisis'.
10. K. Pilaktos, 'Aspects of the external sector of Greek, Spanish and Portuguese economies: a comparative analysis'.
11. C. Courlet and J. Laganier, 'L'Europe du sud: problèmes de l'intégration européenne et scénarios alternatifs'.

3. From *Étatisme* to Capitalism: The Crisis of Development in Turkey and the Arab World

At the gates of industrialized Europe, in the contemporary period, Turkey and the Arab world are experiencing a decisive transformation of their economies and societies. Since the end of the Second World War and especially since the 1960s, history has accelerated in the region. Change has been piled on change, in sharp contrast to the relative stability that marked the Ottoman empire in the previous three centuries. The demographic balance has been upset and the population has doubled or tripled in a quarter of a century. Urbanization has developed very rapidly and today there are as many city-dwellers as there are rural-dwellers who, only yesterday, formed the vast majority of the population. The rural areas continue to employ vast numbers of men but agriculture is no longer the chief activity. Industrialization has advanced rapidly even though it has not always been based on the same sectors and has not everywhere proceeded from the same logic. And the development of modern infrastructures and services has been even more impressive. Education, formerly reserved for a tiny minority of city-dwellers, has made spectacular advances. Growth, more or less regular for two decades, has led to an improvement in the level of average income, which is expressed in increased consumption of manufactured goods and modern services. Traditional forms of production, such as subsistence agriculture and crafts, have disappeared, and where they survive, they do so at the price of a profound alteration of their structures and functions. Ways of life have been transformed, leading to gradual but irreversible changes in collective and individual behaviour. The old social structures are collapsing, the model of the family and the condition of women are changing. Everywhere, traditional community-type solidarities are retreating before the commoditization of social life and the individualization of modes of existence.

Over and beyond sometimes very marked national peculiarities, the countries of the region have evolved in the framework of a single economic model, in which the state assumes responsibility for the development of capitalism in close relation with the world economic system. This model will be defined as a state capitalism model, in order to bring out both the specific features and the common characteristics of the development processes that are underway in the various countries of the region at present.

This state capitalism first appeared in Turkey in the 1920s. It represents a particular combination of *étatisme* and capitalism, and of autonomous industrialization and integration into the international division of labour. It is found later in the Arab world in its essential features, with only some changes in the relative importance of its elements. In some Arab countries, *étatisme* has been more important than capitalism and industrial development has withstood trends and pressures from the outside. Other countries have experienced the opposite situation.

Turkey differs from the Arab world in having a higher degree of industrialization and a more developed export agriculture. And within the Arab world itself, there are major differences between the more *étatiste* and the more capitalist countries at the level of relations between the state and social forces, the nature of industrial development, the type of agrarian structures and the way in which income is distributed. Nevertheless, the points of similarity are very marked.

Nowhere, not even in Turkey, have agrarian structures and social relations in the countryside been transformed in the direction of a decisive improvement of yields. Agricultural production has not risen in proportion to the needs of a growing population, and this has led to a very high degree of food dependence in the Arab world, and conditions of existence for the majority of the peasantry are precarious.

Furthermore, consumption and investment have everywhere depended largely on transfer revenues from the industrialized countries of the east or west; neither in Turkey nor in the Arab world has economic development yet succeeded in integrating the majority of the population into modern forms of production and work processes. These major common limits to the development of the countries of the region have sustained a latent crisis that became manifest once the recession in the West led to a more or less abrupt reduction of transfer revenues.

The crisis is in itself one further element of convergence, as is also and above all the policy followed in the various countries in reaction to the crisis. All the countries experienced an aggravation of the external deficit, growing indebtedness, inflation and unemployment, although for the oil-producing countries these conditions did not arise for another twelve years after their appearance in the other countries of the region.

Whether the open door policies have been put into effect on the initiative of states or at the bidding of the International Monetary Fund and public or private creditors, they are invariably aimed at rolling back *étatisme* in favour of capitalism and creating or augmenting integration into the international division of labour through promoting industrial exports. Everywhere, state capitalism seems to be evolving rapidly towards a system of open classic capitalism in which growth is managed by private agents, both local and foreign, in the context of an increasingly wider opening-up to international competition.

Turkey occupies a place of its own in the Mediterranean basin. Since Ataturk, it was the aim of the elites to make it wholly European, and geographically it undeniably is in part, even though its territories west of the

Bosphorus had been shrinking inexorably even before the fall of the Ottoman empire. But Europe does not include Asia Minor, which was always the centre of the empire and is today the territorial base of the Turkish nation. And by history as much as by culture, Turkey still belongs to its Asian, oriental and Muslim environment. Although Turkey is a member of the Council of Europe, the OECD and NATO, that the external choices of its political class reflect the aspirations of its public opinion is not certain, and it is difficult to forget how different is Turkey compared to its neighbours in southern Europe. West European leaders see the situation accurately when they reject Turkish membership of the Europe of the 12 because of its authoritarian regime and less advanced economy, but also, though less explicitly, because of its different culture. In fact, Turkey's contemporary history has far more affinities with that of the Arab and Muslim world than is generally admitted, and the 'Kemalist revolution' in particular heralded many of the themes that are to be found, in various contexts, in the strategies of social change pursued by Arab states two or three decades later.

State capitalisms, economic growth and social relations

There is not one, single concept of state capitalism. Lenin, who popularized the concept in his analysis of the 'new economic policy' introduced in the USSR in the early 1920s, used it to mean that sector of the Soviet economy based on commodity relations, in the framework of a political regime in which power is exercised by the Communist Party. Later, the concept was used to describe the public sector in advanced capitalist countries, usually made up of nationalized enterprises, or to define the economic system of these countries in terms of the interpenetration between big industrial and banking concerns and the state apparatus, but in that case it is described as state monopoly capitalism. Finally, authors like Charles Bettelheim have recently suggested interpreting the economic system of the socialist countries in terms of state capitalism. This is not the place to discuss the various uses of the concept, as we shall use a different one.

The role of the state in economic and social change in countries east and south of the Mediterranean is more important than it has ever been in the developed capitalist countries. But the pre-eminent role played by the state here differs from that which it performs in the socialist states, for two fundamental reasons. The first is that the *étatization* of the economy is only partial and is locked into commodity relations rather than replacing them. Resource allocation and income distribution are necessarily based solely on the laws of the market, and public enterprises do not necessarily seek to maximize profits; but consumer goods, capital goods and the labour force are commodities and there is a large private sector whose economic, social and political influence goes well beyond its own sphere of activity. The second reason is that this partial *étatization* occurs in the framework of economies that are largely open to the world capitalist economy, in the sense that economic life is heavily dependent on external outlets, products, techniques and financial flows. We shall, therefore, use the term state capitalism to describe an economic and social system in which *étatization* occurs along with the development of internal commodity relations and the maintenance of a high degree of dependence on the world capitalist economy. Indeed, this definition might also be applied to other countries of the Third World, notably the semi-industrialized countries of Asia and Latin America.

In Turkey and the Arab world, the state has been the principal agent of economic expansion and social change, but that does not mean that it has succeeded in remodelling the economy and society in the way it wants, nor that it has really secured control of their processes of change. To say that the state has been the principal agent of growth suggests that growth proceeds from a state project rather than from the spontaneous play of individual initiatives. Of course, it may be thought that there is nothing original about that and that everywhere the transition from the traditional communal or feudal economy to the modern capitalist or socialist economy was variously stimulated or directed by the state. Only the ideologists of 'free trade' and 'freedom of competition' can believe that change in the structures of the economy and society is simply the result of freedom of enterprise, the laws of the market or the mechanisms of comparative advantage. Napoleon III's France, Bismarck's Germany or Meiji's Japan are recent historical examples of situations in which development in a country proceeds as much from a voluntaristic policy by the state as from objective social factors. Even in the United States and Great Britain, countries traditionally described as liberal, the state exercised key responsibilities in the growth of the capitalist economy, over and beyond its functions of security, policing and justice. But in the underdeveloped countries of the Mediterranean basin (and this is a trait that they share with other Third World countries, in Asia or Latin America) the management of economic and social development by the state is much more comprehensive than ever it was or is in the advanced capitalist countries. In those Mediterranean basin countries the state is not content to ensure the 'external' conditions of growth, such as management of the currency, external trade or labour, or the development of infrastructure, but intervenes directly in investment and production and not only in sectors facing a crisis or neglected by private capital.

In the countries of the region, either directly or through its enterprises, the state ensures a very high proportion of investment, as high as 40% to 70% in countries as little '*étatized*' as Morocco and Tunisia. In them all, the state provides a very high proportion of industrial production, from 40% in Turkey to 60% in Egypt, and is the principal promoter of modern services, transport, distribution, communications, insurance, banks, education and health. It often has a monopoly on the exploitation of mines and hydrocarbons, and sometimes a large agricultural domain. Through control of foreign exchange transactions, it centralizes the allocation of convertible foreign exchange resources, as it does sometimes that of domestic monetary resources. Finally, the state largely determines the distribution of income and the structure of expenditure, both because of its enormous weight in the economy and because of its fiscal, budgetary, monetary and foreign exchange policies.

Turkey took up state capitalism earlier than the Arab states, for well known historical reasons, and study of the Turkish model of *étatization* may throw some light on the nature of the development processes of the whole region in the contemporary period.

The Turkish model of *Étatization*

State capitalism developed in Turkey between the two world wars, although its main features had been adopted in the 1920s by the Ataturk regime, which actually drew part of its inspiration from the young Soviet republic. The Turkish model of *étatization*, whose purpose was to initiate the construction of a national economy, was based on the principle of economic nationalism, breaking with the cosmopolitanism of the defunct Ottoman empire. This principle was first applied to external trade, with the introduction of rigorous protectionism and foreign exchange controls, although this protectionism was based on customs tariffs and quantitative restrictions rather than on a state monopoly on imports. At the same time, a degree of control over the domestic market was introduced, through the fixing of industrial prices and agricultural price support.

But economic nationalism came to be identified more closely with direct state intervention in the country's economic activity. An ambitious programme of public investment in industry and mining was launched in the 1920s, while at the same time private capital was encouraged to invest in designated sectors. In such a model, the sources of capital accumulation are domestic, since the surplus available for investment comes initially from agriculture. In fact the state succeeded in keeping local agricultural prices, and notably those of food products, at relatively stable levels and thus succeeded in avoiding a rise in nominal wages in the towns. Since, at the same time, protection of the national market made it possible to raise industrial prices, industry's terms of trade markedly improved and significant profits were realized. State industry was not alone in benefiting from this policy since private industry was allowed to develop alongside it. Similarly, in so far as external trade was not nationalized, the state had to share with the commercial bourgeoisie the rents derived from protection against imported industrial goods.

The policy of keeping agriculture's terms of trade down rested on increasing production rather than on growing recourse to imports, à la Ricardo. At the beginning of the last century, the great English economist advocated the importation of cheap agricultural products, rather than producing them at rising costs, in order to avoid food prices and hence nominal wages increasing and profits and expansion falling. But, apart from the fact that Ricardo was chiefly interested in reducing agricultural activity in Great Britain, both in order to free factors of production for expanding industry and to destroy the political power of landlords, he was comparing the respective costs of food production in a country where land was relatively scarce and in 'new countries' endowed with abundant fertile land. In the Turkey of the 1920s and 1930s, recourse to imports of cereals was not possible because world agricultural prices were higher than local prices.

The state capitalism model in Turkey showed an impressive potential for economic growth. In the midst of the world crisis during the 1930s, Gross National Product (GNP) rose in real terms by 7–9% per annum. Industrial production experienced an average annual growth rate of 11.5% during this period, so that its share of GNP doubled. The contraction of Western markets at this time obviously facilitated the replacement of imports by local production, which was part and parcel of the logic of industrial development, and expansion was accompanied by the reduction in relative terms of imports of manufactured goods, to the point where the previous trade deficit was eliminated.

Turkish industry thus experienced its first burst of expansion thanks to a partial withdrawal, part voluntary and part forced, from the international economy. Furthermore, it was at this time that the process of import-substitution industrialization was initiated in the big countries in Latin America, there too as a result of the combination of the disruption of world markets and a 'developmentalist' economic policy.

Turkish state capitalism had two defining features. On the one hand, control of capital accumulation by the state was inseparable from an involvement of national private capital in industrial investment and in internal (commerce and finance) and external (importing) intermediary functions. On the other hand, the expansion of the economy rested on the national market rather than on exports. Public investment in industry complemented rather than competed with private investment, except for a few branches such as textiles and footwear or cement. Private capital invested in gaps left by the public sector, usually through the creation of small and medium enterprises in technically simple activities, but its markets were often provided by state investment, notably in construction and public works.

The process of industrialization in the 1930s is sometimes described in terms of import-substitution to the extent that it was carried out behind tariff barriers, and treated as similar to what, at the same time, was taking place in Brazil, Argentina and even some countries in Asia still under colonial rule. In reality, this process is to be analysed rather as an attempt at autonomous industrialization in the national framework, rather like the German experience at the end of the nineteenth century. Moreover, the Kemalists very early on adopted the themes of 'national economy' drawn from the German historical school, against the local representatives of the Franco-British liberal current who advocated orthodox free-trade positions. And in 1923 a conference held in Izmir by the Kemalist party stressed the need for the state to actively encourage the emergence of a national class of entrepreneurs and to introduce effective protection against foreign competition in order to develop national industry.

From the social and political point of view, Turkish *étatisme* at the time expressed a class alliance between the national bourgeoisie, then largely confined to agriculture and commerce, and the higher state bureaucracy, both equally interested in the development of a national capitalism. But, in the contemporary international conditions, the emergence of an industrial bourgeoisie in a country on the periphery of the world capitalist system implied a challenge to the position allotted to that country in the international division of labour. In other words, the extension of the social area of capitalism and the takeover of industry by the national bourgeoisie went against foreign interests and called for an autonomous industrial development. The growth of national capitalism was subordinated to the adoption of an industrialization policy that turned its back on the trends and pressures of the international division of labour. In short, capitalism, industrialization and nationalism were synonymous.

Thus, from the 1920s, the Turkish state, acting in a nationalist perspective, strongly encouraged the formation of a national bourgeoisie. First, through the creation of conditions favourable to the individual enrichment of its agents, including its leaders, Ataturk himself being a large shareholder in the Business Bank, nicknamed at the time the 'politicians' bank', which acted as the main pressure group for private interests during the 1920s. Then, by promoting the formation of a commercial and agrarian bourgeoisie. This was built up on a comprador and speculative base by taking the place of the old cosmopolitan oligarchy of the empire. The first fortunes were amassed in the exercise of distribution monopolies handed out by the state, and in property and land speculation. And when the state embarked on an ambitious programme of industrialization in the 1930s, its aim was the formation of a class of entrepreneurs capable of taking over after a period of apprenticeship.

Although the bureaucracy enjoyed wide autonomy with regard to the various fractions of the private bourgeoisie, its permanent aim remained the development of a national capitalism rather than that of a state system. The weight of the public sector and the nature of the policies pursued naturally reflected the relations of force between the bureaucracy and the national bourgeoisie, and the compromises established at a given time between their particular interests. But they also depended on the local and international conjucture and are not necessarily to be analysed in terms of an opposition between the capitalist system and the *étatiste* system. That at a particular stage of development in the economic life of the country there should be less of a role for the state did not in the least mean that state capitalism had been abandoned as a system in favour of a more classical capitalism managed by private agents. The bureaucracy appeared to be acting as the 'enlightened' representative of the whole national bourgeoisie, over and beyond the interests of its various fractions, even if it was only as the potential representative as long as the national bourgeoisie remained insufficiently developed. And while the bureaucracy might oppose this or that fraction of the private bourgeoisie, it did not express an alternative social project to that of the development of national capitalism.

And yet, its internal dynamic made state capitalism an unstable model. Between the state and private capital, relations were in fact very complex. On the one hand, the state closely supervised the private sector which was assigned a well-defined role in the process of economic growth. But, on the other hand, the state encouraged the accumulation of private capital, through the enrichment, licit or illicit, of its cadres, the granting of contracts and public credits, price discrimination, to mention only the principal channels through which the private sector was associated with the mechanisms of capturing the surplus.

Below a certain level of development, private capital cannot assert itself without support by the state in many forms, and without the opportunities for investment created by the expansion of the public sector. But once this threshold is crossed, it attempts to renegotiate its place in the direction of greater decision-making autonomy and a wider field of action. Its position is obviously stronger if, by establishing links with international capital, it is able to appear as a credible investor in the most important sectors of the economy. That is why the possibility of joining with foreign companies or getting access to the world market for credits and technology is one of its principal demands.

It was at the end of the 1940s that the industrial and commercial bourgeoisie in Turkey formulated demands for the denationalization of industry. But when the Democratic Party, representing private interests, came to power in 1950, it was quite unable to sell a single public enterprise, despite its commitments and despite the conversion of some state companies into joint-stock companies, which remained in the hands of the Treasury. Yet, from a technical point of view, a large fraction of the public sector was within the scope of private capital, since two-thirds of the production of consumer goods was coming from this sector at the time. The failure of the attempt to privatize industry is partly to be explained by the more ideological than realistic character of the anti-*étatisme* of the Turkish bourgeoisie of the time. The desire to dismantle the public sector had more to do with the attachment of businessmen and industrialists, and beyond them of a large section of public opinion, to a sort of mythology of freedom of enterprise, than with the objective needs of extending the field of action of private capital. Moreover, this phenomenon was frequently to be found in many countries later, especially during the 1980s, which have seen the resurgence of the influence of the themes of economic liberalism. Ideology does not explain everything, however, and if the privatization of state industry failed, it was also because of quite objective factors. Privatization was in fact only one of many demands of the Turkish bourgeoisie, which had also called for and obtained the liberalization of external trade. And, in the context of a strong expansion of agriculture and the increase in US aid, this liberalization had made commercial activity more profitable than industrial investment.

At the same time, public credits, generally long-term, low-interest credits, granted to private businesses rose sharply during the 1950s, and mixed state–private companies, bringing together private capital, the public sector and foreign companies were created. Financed by public funds, protected from external competition and enjoying hidden subsidies on capital imports as a result of the over-valuation of the Turkish pound, the private sector experienced rapid growth over the next two decades. The state sector gradually came to specialize in the production of capital and intermediate goods, the prices of which were kept at a lower level than those of other industrial prices, and the resulting deficit was covered by the Central Bank or the Treasury.

Table 3.1

Share of the state sector in industrial production

(%)

	Non-durable consumer goods	Durable consumer goods	Capital goods	Intermediate goods
1950	52	—— 73.5 ——		36.5
1963	35.5	4	51.5	79
1980	27	0.2	30.5	55

Source: Boratav, K., 'The role of the state in the development of capitalism in Turkey'.

In 1950, non-durable consumer goods represented four-fifths of public sector production, whereas in 1980, intermediate goods represented two-thirds of it. In the intervening period, the production of consumer durables had risen to the point where it represented one-tenth of the value of industrial production and the state's share in it was insignificant. The growing specialization of the public sector in intermediate goods was, however, accompanied by an expansion of private investment in all sectors. The state's share in industrial production, in fact, fell from around 50% in 1950 to 36% in 1980. But this relative expansion of private industry does not mean that the state has lost its leading role in the process of industrialization. The branches in which the state has gradually come to specialize provide most of the basic industrial products or correspond to the technically more complex activities which have knock-on effects on every productive sector.

State capitalism thus contained a 'natural' tendency to the development of private capital and the industrial bourgeoisie. But the relationship between private capital and the state was not a linear relationship, in the sense that the state would gradually exit, to the benefit of the private sector, from the area of accumulation which it had itself built up. Despite the close links binding it to the various fractions of the private bourgeoisie, the state bureaucracy acted as if it represented a national bourgeoisie which had constantly to be created and strengthened, and whose limits at any given time did not necessarily coincide with those of the class of entrepreneurs. In addition, as the model developed, this bureaucracy acquired an increasingly clear awareness of its own identity and position, and despite its historical solidarity with the private bourgeoisie sought to manipulate those state mechanisms responsible for major decisions about the direction and management of the national economy to act in line with its own interests.

Obviously, as the private sector developed, and an industrial bourgeoisie began to assert itself on the economic and political stage, this power to manipulate encountered more and more limits, and it became increasingly difficult for the bureaucracy to arbitrate between its specific interests and the demands of the development of national capitalism. The management of contradictions between the various ruling fractions became more problematic, especially as the balance established between these fractions was extended into a wider social compromise negotiated with certain sectors of the working class and the peasantry.

In the late 1950s, the realization of superprofits by public, and above all private enterprises in industry as a result of protectionism and the over-valuation of the currency made it possible to satisfy workers' wage demands. In addition, the Turkish state introduced social legislation that was relatively advanced for a Third World country and promoted wage increases while pursuing a policy of over-employment in the public sector. Finally, during the 1960s, the working class took advantage of more favourable political conditions, notably recognition of the right to strike, to organize, especially as its numbers had risen considerably: from 1.3 million in 1963 to 2.7 million in 1977. Real wages in industry thus increased virtually every year, rising by 75% between 1963 and 1976. In addition, the establishment of a parliamentary system led to competition between the parties of the bourgeoisie for the votes of the peasantry, which made up the majority of the country; and this competition compelled every government to guarantee the income of landowners and farmers through agricultural price support. Such a policy broke with the strategy pursued over previous decades to mobilize the agricultural surplus with a view to industrialization. From the 1960s, agriculture's terms of trade underwent a decided improvement, rising from an index of 100 in 1960 to 140 in 1976, especially as agricultural inputs, such as tractors, fuel, fertilizers and seeds, continued to be subsidized; this improvement benefited broad sectors of the peasantry. Despite the determination of governments not to alienate the small and middle peasantry, however, the interests of the big landowners were at no time threatened, and agrarian reform has never appeared in the programme of the ruling parties since the end of the Second World War.

The *de facto* compromise established between the economic interests of the bureaucracy and the private bourgeoisie and those of large fractions of the working class and the peasantry obviously lasted as long as there was no conflict between the increase in the incomes of the popular classes and the profits collected by the dominant classes. Indeed, despite the substantial increase in real wages, income distribution in industry remained virtually unchanged during the 1960s and 1970s.

Long before this social compromise entered a crisis in the late 1970s, it was already far from benefiting the whole population. In industry and the urban economy in general, as in the countryside, major social strata remained outside the redistribution of the fruits of growth. The smallholder peasantry, which lacked the means to extract surpluses for the market, derived no benefit from the improvement in agriculture's terms of trade. On the contrary, it continued to suffer the effects of social changes in the countryside and to fuel the drift from the rural areas and from emigration. In the towns, large numbers of petty employees of the administration and services saw their living conditions deteriorate; and the number of urban unemployed who received no unemployment benefit, rose sharply throughout the period, from a little under one million in 1963 to over three million in 1980.

That is why it is difficult to follow Boratav when he defines the economic system of the 1960s as a populist system, based on a convergence of interests between the ruling classes and a section of the popular classes. The central argument is that the working classes were able, thanks to the parliamentary system, to influence political decision-making and to participate in the benefits of growth without being integrated into the system of power. But according to such a definition, Japan or the United States would be classified as populist systems since in those countries the popular classes remain outside political power but share in the fruits of expansion. In Turkey, the social space in which the gains of development are redistributed is still too narrow for it to be possible to say that the working class and workers in general have been integrated economically. In addition, a social system cannot be characterized solely by its mode of income distribution. The position of the state *vis-à-vis* the various strata and classes of society is, from this point of view, probably a more important criterion. But, the social bloc that dominates the state in Turkey is limited to the higher bureaucracy and the various industrial, commercial and agrarian fractions of the bourgeoisie, thus excluding not only the proletariat, the poor peasantry and the marginalized urban strata, but also a large part of the middle classes in the towns and the countryside. It has been said that when industrialization began, a national bourgeoisie had already been formed in agriculture, commerce and speculation, and the advancement of the old middle classes through state policy was already well underway. The Turkish state, from the very beginning, acted as the historical representative of national capital and the midwife of the local bourgeoisie, by virtue of both the social origin of its leaders and their ideology. In a populist system, the basis of social recruitment of the ruling cadres of the state is wider and more heterogeneous and their ideology is more ambiguous. This means that the state did not function to the benefit of a determinate social class, even a potential one, and the orientation of its policies reflected a much higher degree of autonomy with regard to the interests of the various social strata. Finally, the Turkish state acted on the structures of the economy and society essentially because it was developing capitalism as a social system, whereas in the populist system the state is in itself a major instrument in class formation.

Industrialization, distribution and the international division of labour

The social compromise established at a particular point in time was closely tied up with the country's relationship to the international economy. In an initial phase, the state organized industrialization against the spontaneous tendencies of the international division of labour, which would have confined Turkey to a role of exporter of agricultural products. This phase in the 1930s and 1940s corresponded at the same time to a refocusing of economic activity on the domestic market and a reduction in the relative share of external trade in the national product. The ruling alliance was then dominated by the bureaucracy, the commercial bourgeoisie and the incipient stratum of entrepreneurs, who were organizing industrial development on the basis of the exploitation both of the workers in industry and the peasantry. It is true that this restrictive alliance was facilitated by the absence of a parliamentary regime.

The second phase, from the 1950s to the end of the 1970s saw the acceleration of industrialization in a protected domestic market, but in the framework of a permanent deficit in the external trade balance, compensated by rents in foreign exchange. These rents, which came from US military and economic aid, remittances from migrants, plus receipts from tourism, made it possible to broaden the social compromise of the previous period to the middle and upper strata of the peasantry and workers in the public sector and big industrial enterprises, by enabling them to share in the fruits of expansion.

Finally, the third phase, which began with the 1980s, is marked by the reorientation of industrial development towards external markets in the framework of relatively balanced trade flows. The social compromise has been sharply modified, in so far as the development of industrial exports in the medium term implies both the restriction of domestic consumption and the reduction of production costs, that is, a drastic decline in wages and peasant incomes, at least for as long as substantial advances in production and labour productivity in industry have not been achieved.

The social compromise established around income distribution changed, in that, from one phase to the next, neither the sources of accumulation nor the motive force of expansion were the same. An auto-centred industrialization, based on mobilizing the agricultural surplus and directed towards the domestic market for consumer goods, was followed by growth partly financed by external rents and stimulated by local demand for durable goods, which in turn has given way to an expansion fuelled by profits from industry and sustained by exports. In any event, the Turkish economy continues to be subordinated to the logic of state capitalism, such as we have defined it, that is, to a logic of incomplete *étatization* that develops a national capitalism and manages its relations with world capitalism.

Because the industrialization process in this model is managed by the state, foreign capital cannot have a hegemonic position and its relations with the public sector cover complex relations of opposition and co-operation. The state was firmly opposed to foreign capital during the first phase of industrialization, when it was investing against the spontaneous tendencies of the international division of labour. But it desperately needed the technology that foreign firms possessed and the equipment that they made, and even the credits that they could mobilize. For foreign capital, industrialization was a threat to its traditional outlets, but it was also a new opportunity for profits. In fact, direct investment by big Western firms in Turkey remained limited, compared to other regions of the Third World. Turkey is, of course, poor in mineral resources and agricultural raw materials and also the level of wages is relatively high, yet the size and protection of the Turkish market constitute undeniable attractions. But, in so far as it has been able to enjoy external rents, the state has given only slight encouragement to the inflow of foreign capital, at least until the present period. The stock of foreign investment in the country is extremely modest relative to total investment. Although the real amount of foreign capital is probably higher, since its investments are usually made in association with local capital, public or private, and it is interested in technologically complex sectors.

If one were to define the essence of the Turkish model, the three key terms would be state, capitalism and economic nationalism. In the context of the 1930s industrial development in an agrarian country implied a break with the objective tendencies of international specialization, and opposition to all foreign capital. Industrialization, respecting the essential rules of capitalism – private ownership of the means of production, wage labour and production for profit – then necessarily had a national character, and not only because it was managed by the state. It is in this sense that the industrial bourgeoisie, actual or potential, in an agrarian country was at the time a national bourgeoisie whose interests were in contradiction with those of imperialism. When, a few decades later, the Turkish model opened up to the world capitalist system, first by increasing its imports, then by developing its industrial exports, it had meanwhile made possible the formation of an indigenous bourgeoisie, one that was certainly fractionated but still economically dominant. Reintegration into the international division of labour thus occurred on new bases, the country having succeeded in securing for itself a better place, after a period of withdrawal. And while it disrupted the coherence of the national productive system, the international specialization of Turkish industry does not mean the abdication by the local dominant class of its power of economic decision-making.

State capitalism and development in the Arab world

The 1960s and 1970s, during which most Arab countries emerged on to the world economic stage, were marked by a quite different context. The international division of labour had evolved in such a way as to make it possible to relocate certain industrial activities in certain Third World countries. And the development of industry in agrarian countries no longer implies, as it did in the past, the infringement of the rules of international specialization, or at least not in all cases. Sometimes it was foreign capital itself that organized the industrial relocation, sometimes it was the local states that forced it on them by playing on the competition that set then against one another. But usually, the expansion of industry in the Third World reflects a larger movement to reallocate industrial work within the capitalist world. It can thus be effected without any necessity for even a temporary withdrawal from the international economy or for making arrangements for autonomous growth in a closed economy. From this point of view, no longer are the interests of the dominant industrialist groups, whether state bureaucracy or emerging industrial bourgeoisie, necessarily opposed to those of international big capital, although they may be to those of investors less well equipped to face competition on the world market. In these conditions, a process of industrialization effected in a capitalist framework creates an appendage to the international economy rather than the base or the means for building a national economy. Autonomous industrial development not, of course, in the sense of autarky but of control of decision-making power, is now only possible if it turns its back, at least for a time, on the logic of capitalism, and this means that local capitalist tendencies have to be repressed and international specialization rejected.

Turkey's contemporary experience is original in more than one sense, and many aspects of its development model give it a special position in the region. But because it belongs to the same cultural area as the Arab world, which indeed it dominated for several centuries, Turkey has often shared with it the same social or ideological movements. And while, since the fall of the Ottoman empire, it has loosened its links with its neighbours to the south, Kemalist nationalism was a strong influence on modern Arab nationalism, although the latter also drew on other sources.

The economic and social evolution of the Arab world since the 1960s belongs to the state capitalism model such as we have defined it, that is, a dual process of strengthening the state and the market in the framework of close integration into the capitalist world economy. Compared to the Turkish model, two principal variants may be distinguished, according to the degree of *étatization* of the economy (and society) and the type of integration into the world capitalist system.

The first case is that of such countries as Morocco, Tunisia, Saudi Arabia and the Gulf countries, in which the state manages the conditions for the emergence of national capital in the narrow framework laid down by the international division of labour. The second is that of countries such as Egypt in the 1960s, or Algeria and Syria in the 1970s, in which a much more active state attempts to build an autonomous public economy, disregarding the international division of labour. The social basis of the state, the field open to the development of private capital, the place of foreign capital and the logic of investment are different in the two cases, as are the structure of production, the ways in which the currency is managed, income distribution and the social demand profile. It could be said that the two cases differ principally in the economic policies pursued and the nature of the social processes that condition them through the state. They make it possible to set out the broad outline of the principal lines of economic and social evolution in the Arab world, though it would be pointless to seek an accurate illustration of them in a particular country.

In addition, the second case, which is more *étatiste*, has proved to be historically less stable than the first, since *étatization* tends to be followed by opening up, both to national and foreign private capital and to the international division of labour.

For reasons of convenience, we shall call the first case, in which the capitalist logic wins out over the *étatiste* logic, 'open state capitalism' and the latter 'populist state capitalism'.

Open state capitalism

In Tunisia and Morocco, the leadership of the national movement by the middle strata in the towns led to the colonial model of specialization in agriculture and mining being called into question. The local bourgeoisie was primarily mercantile and agrarian and was not in a position rapidly to replace colonial interests in modern spheres of activity and notably in industry. The state thus found itself obliged to undertake a number of tasks itself – nationalization of the mining sector, development of infrastructure, (partial) nationalization of the banking system, and finally investment in industry. From the beginning, however, its action was conceived both as a (temporary) substitute for private initiative and as the means of ensuring the development of national private capital. The state did not engage in the economy in opposition to the local bourgeoisie, nor did it propose an alternative economic and social project to that of private capitalism. As early as 1957 in Morocco, the government set up a *Bureau d'études et de participations industrielles* (Office for industrial studies and shareholding) whose function was to build up investment in industry in association with national and foreign private capital. In 1959, it set up a *Banque Nationale pour le Développement Economique* (National Economic Development Bank), in which private capital, both Moroccan and European, held the majority interest; its activity consisted in financing private investment, both Moroccan and foreign, in profitable short-term operations, such as assembly, textiles, food processing or phosphate processing. Public investment developed rapidly in these same sectors, pending the time when national capital could take over. By the late 1970s the state was involved directly, through 460 public enterprises, in wide areas of agriculture, industry, tourism, mining, banking and distribution. Meanwhile, however, Moroccan private investment, which used to represent a negligible proportion of total investment, now represented 45%.

In Tunisia too, while the state took the initiative of developing small-scale steel-making, activities inherited from the colonial period, such as cement manufacturing, small-scale metal-working and textiles, were left in the hands of French interests, and national private investment was actively encouraged. During the 1960s, the Tunisian private sector was responsible for approximately only one-fifth of crude capital formation, largely concentrated in agriculture and tourism. But during the following decade, it came to dominate wide areas of industry, notably textiles (81% of total investment as against 28% during the 1960s), agricultural food processing (60% as against 20%) and electrical and mechanical assembly (53% as against 9%).

In both countries, the role of the state in the development of industry declined as industry attracted private capital and a class of indigenous entrepreneurs emerged, as had happened in Turkey a few decades previously. For the urban middle classes that had taken over senior positions in the state, acceptance of the institutions and values of capitalism was fuelled by the idea that the removal of the colonial rules of the game would make possible their own advancement in the framework of capitalist relations of production.

Contrary to what happened in Turkey in the 1930s, however, state intervention in economic life here was not aimed at building an autonomous industrial base, in opposition to both the strategies of foreign capital and the spontaneous tendencies of the international division of labour. The state encouraged the formation of a bourgeoisie of entrepreneurs, but on the basis of a limited and fragmentary industrialization, in line with the country's comparative advantage, and in association with foreign capital.

For the most part, public industrial investment and, *a fortiori*, of course, private investment was directed towards activities associated with the consumption demands of minority strata of the population and external demand. That it was oriented towards import substitution or export promotion is less important than the fact of its total subordination to the international division of labour and the strategies of foreign capital. Production was largely centred on the food processing industries and textiles, which, in Tunisia, for example, represent over half the value-added in industry, building materials, mining and phosphate processing and electrical and electronic assembly activities. Industrial activity is limited either to the manufacture of non-durable consumer goods, for the domestic market, or to the assembly of semi-durables (clothes) or durables from imported components, for the external market. The international competitiveness of production rests essentially on the low level of the cost of labour. There has been no significant development of basic industries, apart from Tunisian small-scale metallurgy and phosphates in Morocco, and neither the private sector nor the state has sought to increase the local linkage of industrial production: it was more profitable to import the components, the semi-finished products and capital goods rather than make them on the spot.

The public industrial sector generally operates at a loss but not because it produces intermediate and capital goods below cost for the benefit of the private sector, as happened in Turkey between the 1930s and 1960s. Its deficit rather reflects irrational management in the sense that its economic objectives are compromised by the operating constraints brought about by interventions by politicians. In addition, while the public sector here is not called upon to play a role in sustaining global growth, it does not function as a capitalist enterprise since its management is partly linked to procedures for distributing political and material advantages within the ruling elite, and the incomes that it secures for the people and groups that run it are taken from the public receipts rather than from the market.

Such a fragmentary industrialization and one so dependent on the world economy is obviously unlikely to lay down the bases for an autonomous development of the national economy, especially as it is directly subject to the intervention of foreign capital. In Tunisia, direct foreign investment increased considerably in the course of the 1970s to the point at which it represented one-third of capital inflows. While it was mostly invested in petroleum, it was also important in industry, relative to Tunisian capital. For example, 85% of the 352 projects realized in the framework of the April 1972 law on the promotion of industrial exports were the work of foreign enterprises, generally small and medium firms in Western Europe, acting alone or in association with local interests. In Morocco, too, the enlargement of the private industrial sector appears to be closely linked to the growing penetration of foreign capital. And, despite the restrictions envisaged in the laws, 'Moroccanization' has not prevented the formation of associations between local interests and foreign firms. Moreover, the 1982 investment code removed these restrictions and moved in the direction of total liberalism with regard to private capital in which the foreign element predominated.

Although the availability of a significant petroleum rent and the particular nature of their social and political structures put the Gulf countries and Saudi Arabia in a category of their own, the way in which they have evolved economically broadly obeys the same logic. There, as in Morocco or Tunisia, the state has endeavoured to develop a modern local bourgeoisie in the framework of the international division of labour and in association with foreign capital. But here, it is rather a financial bourgeoisie, whose economic position is based on access to petroleum rent rather than a bourgeoisie of industrial entrepreneurs. Here, industrialization is even more strongly integrated into the international economy, since it is centred on primary processing of hydrocarbons and intermediate goods industries for export, although in the case of Saudi Arabia, the local market plays a not insignificant role. Petroleum refining, petrochemicals, steel-making, the aluminium industry and building materials are developing there very fast, in the framework of mixed enterprises bringing together the state and foreign capital, and with the direct participation of big Western and Japanese firms in the construction – and the management – of large-scale factories utilizing the latest technologies. Most Gulf countries have giant refineries, which have enabled them to increase substantially their share of the international market in refined products, to the point where European countries are currently developing protectionist reactions. Saudi Arabia possesses enormous capacities for producing steel, petrochemicals and cement, which are way beyond the capacity of its domestic market to absorb. Comparative advantage here is based on the availability of energy at very low or virtually zero cost, but industrial specialization in high-energy content intermediate goods is also a dependent specialization. The advantage represented in terms of industrial learning by the production of intermediate goods such as steel or plastics compared to the production of non-durable consumer goods or simple assembly is made up for by the greater dependence on a handful of big Western and Japanese firms which own the technologies used. In addition, the direct management of most industrial complexes by foreign firms participating in joint venture projects is not specially favourable to the accumulation of industrial experience by local entrepreneurs, whether private or public, although the massive training of technical cadres in Anglo-American universities may well make possible an eventual take-over from foreign technicians and managers. If the Gulf countries and Saudi Arabia occupy an apparently more solid position within the world capitalist system than that of Morocco or Tunisia, it is obviously because of their energy wealth and their financial power and not because of their new industrial specialization.

Compared with the Turkish experience of the 1930s to 1950s, the contemporary evolution of the Arab countries pursuing 'open state capitalism' has no element of economic nationalism. The development of the public sector and the promotion of the private sector by the state have occurred without breaking with either the international division of labour or foreign capital, although in the countries which were European colonies, the 'pacte colonial' has been called into question. For that there is one fundamental reason: in current conditions, the development of local industrial capitalism is both possible within the world capitalist system and incompatible with the building of an autonomous national economy. The economic system of world capitalism now accepts the siting of some industrial activities in some Third World countries, whether it is directly ensured by transnational companies or left to the initiative of the state. But precisely because local capitalism can flourish in these conditions, its rationality turns it away from any ideas of autonomous industrialization. 'Open state capitalism' looks in fact like a pseudo-state capitalism, since neither the state nor private capital has any strategy of building a national industrial economy. It is a pseudo-state capitalism dominated by the compelling tendencies of the international division of labour, or even directly by big international capital. When the state develops local capital, it is in reality contributing more to the emergence of a further link in the chain of international capital than to that of an agent directing national development. When the higher bureaucracy actively encourages the formation of an industrial bourgeoisie, often in its own ranks, it gives rise to groups that are auxiliaries to the foreign bourgeoisie rather than to a class capable of managing the growth of the economy. It is all as if the state in this model was acting as the principal promoter of the dependent integration of the country into the world economic system, on more rational bases than during the colonial period, given that such integration rests on the association of local groups which are minorities but in no way insubstantial.

Other Arab countries have had a different experience during the last two decades, marked by economic nationalism, in which *étatisme* determines the logic of economic and social development far more, without, however, breaking either with the laws of capitalism domestically or with the system of the world economy.

Populist state capitalism

As a result of particular historical conditions, in Egypt in the 1960s and in Algeria and Syria in the 1970s, the state represented a broad class alliance, centred on the urban middle strata, but including elements of the popular strata in the towns and the countryside. Unlike the Turkish model, the demands for compromise among these various social strata made it impossible for the state to be turned into an instrument for the benefit of the bourgeoisie. Because it had broken with the old political structures, state power was largely autonomous *vis-à-vis* local bourgeoisies and middle classes, even if it was dominated by radical groups in the petty bourgeoisie.

From the start, the state undertook to limit the field of expansion of national private capital and the influence of foreign capital on the country's economy. It nationalized settler lands but they were not handed over to well-off peasants or urban traders. It set up a large public productive sector in agriculture, as it did in mining, banks and industry, on the basis of the nationalization of foreign interests. It launched an ambitious programme of public investment notably in the capital and intermediate goods industries from which private investment was generally excluded. Private capital, which was also barred from financial intermediary roles, could invest only in sectors defined as non-strategic, such as the non-durable consumer goods industries, small construction materials, milling, and so on. As for foreign capital, it was accepted only as a minority partner of the state and in activities on the periphery of the productive system strictly so-called, such as engineering and construction. Finally, a state monopoly on external trade was instituted, which involved a close control of the private sector's possibilities of expansion.

As in Turkey, but without an industrial bourgeoisie playing an active role, this model showed a high potential for industrialization. Industrial development there was centred on the domestic market and rested on sectors (capital and intermediate goods) that involved a qualitative leap in the economy's global production function and a rejection of specialization in terms of comparative advantage. There was also more learning of techniques by workers and managers, in so far as the management of factories with complex modern technology was assured by the public sector rather than by foreign firms.

In the populist model, two underlying forces are at work. The state is taken over by the middle strata and elements of the popular strata (petty officials, employees in the service sector, artisans, traders, small peasants, and so on) but *étatization* leads to the formation of new social strata which replace or overlie the old ones. The development of state interventionism causes the state apparatus itself to be the mainstay of the formation of new social strata: employees in the administration and the public services, teachers, technicians, workers and cadres in the public industrial and agricultural sector, managers, and so on. Among these new social strata, those that occupy the commanding heights of the state apparatus gradually transform themselves into new ruling strata.

In addition, because *étatization* of the economy is only partial, it also engenders the formation of new social strata outside the state apparatus and the parastatals. The fact is, that state intervention gives rise to new opportunities for profit and investment in a large number of activities which are left to private initiative. The construction, small-scale services, industry, distribution and leisure sectors enjoy very high demand, maintained by the growth of public contracts and incomes distributed by the state. Investment in these sectors attracts above all middle traders and peasants, who possess the necessary funds, experience or contacts, but people of modest origins can also take advantage of opportunities for speculation created by shortages and planning dysfunctions. Thus, new privileged social categories come into being, entrepreneurs in industry and services, rich peasants, the liberal professions, not to mention the cadres in the state apparatus who become involved in the business world.

In short, these social changes result from the combined effect of the development of *étatisme* and the fact that it is not developed enough. If the state promotes the formation of new privileged strata within and outside itself, it is because the logic of its expansion is not unequivocal. It is not a completely *étatiste* logic, totally at odds with that of the market and capitalism. Nor is it the simple expression of a capitalist or commercial rationality. It is a hybrid logic, half-*étatiste* and half-commercial. Resources are allocated and incomes and statuses are distributed according to both the criteria of an *étatisme* more or less distorted by commodity relations and the rules of a market more or less modified by state intervention. If *étatisme* and the market develop simultaneously, the social space of their development is in fact a single space: the logic of each conditions and affects the logic of the other. And if this ambivalence of the dominant economic rationality enables the middle classes to base their rise on the simultaneous extension of *étatisme* and commodity relations, it is to be explained by the determination of the middle classes to keep all the channels of their advancement open.

Finally, the state is not above society, despite its great autonomy in relation to social forces, and the way it functions reflects more or less subtly the traditional sociological vectors and divisions. The bureaucracy that develops with the extension of the role of the state into economic life is not a Weberian bureaucracy but a group whose bonds with society remain very much alive. It does indeed aspire to rationalize social activities but it is also imbued by traditional clan, regional, ethnic or confessional solidarities, especially in countries where the state tradition is only minimally developed.

On the economic level, while the populist model has led to significant industrial development and made possible the establishment of a technical infrastructure, it has, after what was historically a relatively short phase, run up against serious limits. It has no more escaped food and financial dependence than the open capitalism model nor has it demonstrated a superior capacity to settle the problem of employment, but, in addition, it has encountered specific difficulties. While industrial linkage is higher than elsewhere, it is far from satisfactory, and the economy remains heavily dependent on the outside world for the capital goods it needs. Industrialization developed exclusively on the basis of foreign technology and no consistent policy to promote national technological capacities was implemented, although there were some efforts in that direction in particular sectors. Certainly, the state domain was very large in production, distribution, finance and importing, but conflicts within the state apparatus between the various decision-making centres often led to irrational management of state agencies and enterprises. The public sector in agriculture and industry was often inefficient, both because of constraints and difficulties arising from its environment and the organizational problems within it. This model thus suffered from an enormous waste of labour and capital. In addition, planning dysfunctions have been constantly aggravated by the presence of the private sector and the development of commodity relations within the public sector. While private capital was closely supervised by the state, it enjoyed concealed subsidies for the intermediate goods sold by the public sector at controlled prices, public contracts, notably in construction, and even skilled workers trained by state enterprises. Moreover, the rationing of private investment by the state aimed at channelling it towards determinate sectors paradoxically resulted in stabilizing oligopolistic positions, and thus reducing competition, discouraging falls in prices and costs and the quest for quality, and promoting the formation of irregular incomes which would fuel luxury consumption and currency speculation.

As the economy advanced, society became more complex and the positions of the various strata of the population became differentiated: it became more difficult for the state to reconcile their interests, especially when the potential of the model was reaching its limits. The bourgeoisie, which had come into being in the meanwhile in industry, services and commerce, was now more conscious of its weight and sought to renegotiate its position in the system just when the conflicts within the state apparatus between the various bureaucratic and technocratic fractions over the orientation of economic policy were becoming sharper. The Egyptian *infitah* marked the triumph of the fractions of the state bureaucracy favourable to an opening up of the system to local and foreign capitalism. In Syria and Algeria, while the opening is developing in a particular way and while a new integration into the international division of labour is also on the agenda, the outcome of the restructuring underway is much more uncertain.

To a greater or lesser extent a large peasantry, and rural activities dominate all the countries to the east and south of the Mediterranean. We have seen how decisive has been the role of agriculture in the economic development of Turkey, first as a source of capital accumulation and then as a market for industry and a social base for political power. State capitalism in Turkey failed to modernize agriculture, which still employs half the economically active population, but succeeded in developing an efficient agrarian capitalism. In fact Turkey is currently the only country in the region to have a grain surplus – whereas the story of agriculture in the Arab countries is one of a double failure: of capitalism and of *étatisme*.

Agriculture, growth and food dependence

In Turkey, social differentiation in the countryside was effected through a classic process of the development of agrarian capitalism during the 1940s and 1950s. The state promoted the expansion of large farms by distributing long-term, low-interest advances to rich peasants. Irrigation and mechanization spread rapidly, along with the growth of agricultural exports and the raising of domestic and external prices. The increase in production on large modernized farms ruined part of the small and middle peasantry, promoting land concentration and proletarianization. Agrarian capitalism did not spread to the whole farming area and some regions of the country continued to be marked by archaic structures. But it did succeed in greatly increasing production and yields so that agriculture sustained industrial expansion, by supplying the towns with food, extending the domestic market and producing surpluses for export. It is true that, compared to its Arab neighbours, Turkey benefited from a number of favourable conditions. First, the absence of a colonial past enabled it to escape the distortion typical of Arab agricultures in which capitalist modernization was often introduced from outside and focused exclusively on export crops. Second, agrarian capitalism developed in Turkey over a long period in a favourable situation marked by rising external prices and relatively low domestic costs. Finally, the natural conditions for exploiting the land are decidedly better than in the Arab world.

Agriculture in the Arab world suffers from unfavourable natural conditions. Desert, steppe, mountains and marshes cover most of the territory, and, except in a few coastal regions, rainfall is inadequate and irregular. Throughout history, intensive working of the land has rarely, except in the case of Egypt, constituted the main economic base of Arab societies, which have rather been organized around pastoralism and trade. Agriculture in the east and the Maghreb, centred on extensive cereal growing, has always been separated from livestock farming and, in addition to unfavourable ecological conditions, there is thus the traditional low skill level of the peasant labour. The low level of yields and productivity and the limited scale of the agricultural surplus were masked by the fact that populations were relatively small and scattered and the majority had very low consumption levels. The upheavals experienced by the societies of the Arab world in the contemporary period, population growth, urbanization, the improvement in average income and changes in patterns of consumption have clearly revealed how poor is the agricultural potential of most countries. The development, over the last two decades, of mechanization and the use of fertilizers and pesticides on much of the arable area have not led to any great improvement in yields.

Irrigation has been developed in a few countries, such as Morocco and Iraq, but for all countries in the region it remains insufficient. Algeria irrigates only 250,000 hectares, or 3% of the cultivable area. Rather than small-scale, more labour-intensive water projects based on the use of underground water and flood waters, states have preferred to build large, and costly, dams and the retention capacity of these is limited by silting.

The level of education and training of the peasantry remains extremely low. The economically active rural population, which includes a large proportion of illiterates, has benefited little from the educational systems set up in the various countries, and technical and vocational training intended for agriculture is extremely inadequate. As for agronomic research, nowhere has it measured up to the technical needs.

Cereals, which are the basic foodstuff in the Arab countries, occupy most of the agricultural land (73% in Algeria, 51% in Syria, 60% in Morocco) but these are grown mainly on non-irrigated areas and suffer heavily from the effects of irregular rainfall. In the Maghreb, biennial fallow–wheat rotation has been maintained, despite its outdatedness and the shortage of land. Almost 40% of the cultivated area is thus rested each year in Algeria. The general fall in the fertility of the soil, whose nutritive elements have not been reconstituted as they have been exploited, is reflected in very low yields per hectare. Apart from Egypt, no Arab country obtains a yield equal to the world average of 2,000 kg; the actual yields obtained range, depending on the productive system, between 500 and 1,000 kg. Furthermore, the fall in soil fertility is not compensated by the use of artificial fertilizers, because degraded soils, suffering from poor and irregular rainfall, have difficulty in absorbing them. The production of cereals in the whole of the Arab world is equivalent to only 62% of that of Canada, 57% of that of France and 10% of that of the USA, and nine-tenths of the arable areas of the Arab world are in Sudan alone.

The new priorities identified during the 1980s, especially in the Maghreb, do not appear to call into question previous policies in soil use, irrigation, training and agronomic research. In general, there is a move towards accentuating mechanization and greater use of chemicals in production, through increased recourse to imports of machinery and fertilizers.

On the whole, agricultural policies have failed to make agriculture the source of capital accumulation in the Arab world. In the non-oil producing countries, the deterioration in the terms of trade imposed on the agricultural sector simply reflected the fall in incomes of the majority of the peasantry.

In Egypt, some agricultural products, such as cotton, a key export product, were marketed solely by co-operatives. Most basic foodstuffs – rice, and onions for example – were subjected to compulsory deliveries to the state. Usually, producer prices were kept at a low level compared both to consumer, and industrial prices. The use of fertilizers and pesticides has been subsidized and the state has made enormous investments in irrigation and drainage; nevertheless, it has been estimated that the manipulation of prices to the disadvantage of agriculture has led to a transfer of 5–7% of total agricultural income to the towns and to industry.

In Tunisia too, the terms of trade moved in a direction unfavourable to agriculture as a result of the state freezing producer prices of the main agricultural products. At constant prices, hard wheat, for example, fell from 4.2 dinars per 100 kg. in 1961 to 2.8 dinars in 1979. Simultaneously, industrial prices, particularly those of agricultural inputs, increased regularly, so that production costs became higher than the prices of the main products. In 1975, the cost of production of wheat per metric ton was 75 dinars: the producer price was only 66 dinars. It has been calculated that, in 1975, the gap between the cost of production and the producer price for cereals, olives, milk and meat corresponded to a levy of 16% of the value of agricultural production. During the 1980s, however, the state has modified its agricultural pricing policy and allowed an improvement in farmers' incomes, aimed at facilitating an increase in production and reducing imports that have become too expensive relative to local prices and costs of production.

In the oil-producing countries, the rent extracted from the export of oil and gas has obviously relieved the state of the need to mobilize an agricultural surplus to finance growth, and, conversely, made possible an improvement in agriculture's terms of trade. In Algeria, producer prices have been increased for all products, although at higher rates for fruits and vegetables and meat, while the prices of industrial inputs have been kept at a level well below their cost of production or their landed import price. The delivery price of a tractor, for example, is much lower than its landed import price, while the national currency is overvalued and the cost of production of fertilizers is twice their sale price. In addition, the public and co-operative sector deficit has been regularly covered by state subsidies, which have made possible a marked increase in agricultural wages that bear no relation to labour productivity.

Food dependence

Every Arab country, except for Sudan, has become a large food importer. Their production now covers little more than half their needs, whereas 20 years ago they were almost self-sufficient. To the traditional imports of coffee, tea, sugar and oil have been added products usually available in the country: cereals above all, but also pulses, potatoes, onions, milk, butter and meat. Traditional exports (wine, olive oil, dates, logs, cotton) have continued and new exports (early vegetables, fruits) have been developed, but the balance of agricultural and food trade has still deteriorated.

In Morocco, between 1960 and 1980, self-sufficiency rates fell from 86% to 72% for cereals and from 61% to 50% for dairy products, while the cover rate of agricultural imports by exports fell from 156% to 70%. The food deficit is much more marked in Tunisia, where the cover rate fell from 150% to 50% over the same period. Local production now accounts for only 55% of cereal needs, and 40% of dairy products. Egypt has to import three-quarters of the wheat it consumes. In Algeria, production now provides only half of consumption (as against 70% in 1970). For essential products, production now virtually plays only a supplementary role, since it is far below imports for cereals, pulses, vegetable oils, milk and dairy products.

The appearance, or the aggravation of the food deficit in the Arab countries, is as a result of a combination of low growth in agricultural production and a rapid increase in food consumption. This increase is due not simply to high population growth, but also to an increase in food consumption, especially in the populist model countries, as a result of urbanization, the expansion of employment and the increase in average income. In Algeria, for example, annual per capita consumption of cereals rose from 147 kg in 1960 to 187 kg in 1980; of vegetable oil from 4.7 kg to 13.3 kg; of meat from 6.6 kg to 15.6 kg; and of sugar from 17 kg to 22 kg. The food deficit thus reflects not only a particular state of production, but also a given level and structure of consumption. Furthermore, a food deficit can emerge despite the existence of agricultural surpluses if these are obtained by cutting down popular consumption.

The development of food dependence during the 1970s was to some extent a deliberate strategy on the part of states, even though it was the short-term inelasticity of production faced with a strong rise in consumption that led to the recourse to imports. Until the 1980s, it was preferred to import food rather than increase local production at higher prices because of poor yields and low agricultural labour productivity. Recourse to cheaper imports in fact made it possible to avoid the wages of permanent workers in the towns rising too high, and to prevent the impoverishment of large sectors of the urban population. But recourse to imports did not relieve the state of the need to support the prices of basic necessities, given that although imported foodstuffs' prices may be lower than their domestic production costs they were, nevertheless, higher than the average income of the urban popular classes; in fact, budgetary subsidies have increased in line with foodstuff imports. In Algeria, for example, almost nine-tenths of subsidies go to cereals and pulses, whose consumer prices have thus been stabilized despite the increase in their production costs. In Tunisia, subsidies rose thirteen-fold in six years. On the other hand, the upward trend of cereal prices on the world market in recent years has reduced the gap between import prices and local costs of production and made it easier for states to contemplate developing local production. Finally, above a certain threshold, food dependence becomes politically intolerable, even in open capitalist countries. This explains why reducing food dependence has featured on the agenda almost everywhere in recent years. But knowing the social and economic conditions that are required for strategies of food self-sufficiency to succeed is another question.

Agrarian reforms and land tenure

In the open state capitalism countries, agrarian reform has been very limited and the concentration of landownership remains very high. In Tunisia, 1.5% of farmers currently own over one-fifth of the land (as against one-quarter in 1960). Only settler lands have been nationalized, and these make up a state domain of some 400,000 hectares. A second agrarian reform was implemented during the 1960s, with the establishment by fiat of co-operatives grouping together poor and landless peasants, but it aroused hostility from those principally concerned and ended in failure.

In Morocco, 0.5% of farmers own 17% of the arable land, which is an even higher degree of concentration than in Tunisia. The Moroccan agrarian reform, also limited to the settler sector, has not greatly affected the structure of land tenure in the country. One million hectares of settler lands were nationalized, of which 300,000 have been sold at preferential prices to peasants, to be set up as co-operatives.

In the 'populist state capitalism' countries, agrarian reforms have naturally been more radical, in that they have affected indigenous landholding. In Egypt, the 1952 reform led to the disintegration of the old agrarian structures. One million *feddans* were nationalized and distributed to some 340,000 families of agricultural labourers and poor and landless peasants. A movement to form co-operatives was launched a few years later, with a view to facilitating the rationalization of farming, but property continued to be individually owned. But it is in Algeria that the agrarian reform went furthest, as was predictable in a country which experienced a profound colonization based on the confiscation of land and where the poor peasantry played a considerable role in the national liberation war. Immediately after independence, settler lands were nationalized and handed over to self-management, although in practice, self-management was limited to the organization of the labour process, the self-managing estates functioning economically as public enterprises managed by the Ministry of Agriculture. In 1971, a second agrarian reform tackled large-scale Algerian-owned property, confiscating the land of absentee landlords, and land above a ceiling of an average of 50 hectares set for private farms. Thus 500,000 hectares were nationalized and distributed to 100,000 landless peasants, in the framework of producer co-operatives. Recently, these co-operatives, whose economic performance was generally mediocre, were dissolved and the lands distributed to co-operators or added to self-managed estates.

Nowhere, however, did the agrarian reforms succeed in altering land tenure structures and social relations in the countryside in a direction favourable to any distinct increase in agricultural production. In the open capitalism countries, agricultural policies have failed to develop agrarian capitalism and rationalize the land and labour markets in the countryside. In the populist countries, the state has established a major public sector, but it has failed to provide proper operating conditions or to make it the driving force in the modernization of agriculture.

Overall, two typical trends are apparent: the consolidation of small-scale family holdings, and the dynamism of 'kulak'-type rich middle peasants. In Tunisia, the numbers of smallholder peasants (with under five hectares) and small peasants (five to 20 hectares) has increased since the 1960s, while the number of permanent agricultural wage labourers has declined; wage labour accounts for about one-tenth of the agricultural labour force. Tunisian agriculture has thus not experienced the classic capitalist process, based on the proletarianization of smallholder peasants and the concentration of land in the hands of an agrarian bourgeoisie, nor has it experienced a democratization of ownership, since it continues to be marked by a very high concentration of ownership. Conversely, larger medium-sized property (50–100 hectares) has spread significantly, in terms of numbers involved and areas covered, at the expense of large estates, which have lost one-third of their area since 1960. The recent evolution of Moroccan agriculture similarly illustrates the consolidation of small family holdings, although the concentration of land and the proletarianization of the smallholder peasantry have gone further there. The number of landless peasants and agricultural labourers has increased immensely since independence, and the number of employers has diminished, which is the sign of a greater flowering of agrarian capitalism. But that has failed to alter the structure of the economically active agricultural population:

the proportion of workers has remained at one-fifth during the last three decades, whereas, over the same period, that of family helpers has risen from 25% to 35%.

In Egypt and Algeria, the agrarian reform destroyed the rural aristocracy of big indigenous landlords and led to a redistribution of land and income in favour of the poorest stratum of the peasantry. But this redistribution was limited by the desire to avoid excessive subdivision of nationalized lands. The one million hectares distributed in Algeria under the agrarian revolution (half of which came from the confiscation of the land of absentee landowners and the limit set on private holdings and half from the development of public lands) benefited 100,000 allocatees grouped into co-operatives, whereas there were 600,000 landless peasants at the beginning of the 1970s, in addition to smallholder peasants. Similarly, in Egypt, the agrarian reform failed to satisfy the ownership aspirations of the majority of landless peasants. The aim of the agrarian reforms was, in fact, as much social and political as economic. The aim was to destroy the political power of the big landowners and limit the rural exodus by improving living conditions in the countryside. But it was also designed to increase agricultural production and yields in order to ensure the food supply for the burgeoning urban population and enlarge the market for industry, and this ruled out the granting of a tiny plot to every landless peasant.

Conversely, while the agrarian reforms eliminated large estates and prevented the development of an agrarian capitalism, the efforts made to encourage small peasants to join production or service co-operatives failed. This meant that there, too, family and smallholder farms were maintained or consolidated, thus giving free rein to new processes of social differentiation which led, as in the open capitalism countries, to a significant growth of the stratum of middle and upper middle peasants, who, according to the area, had between 10 and 50 or 100 hectares, or even smaller plots on irrigated land.

These processes did not derive from the concentration of property ownership, although land transactions played a not insignificant role in Egypt and of course in Morocco and Tunisia, but from the spread of the hiring of equipment and labour and the degree of openness to the market for agricultural products. Mechanization has developed considerably in the private sector, but few peasants own the equipment they use.

In Algeria, over half the private farms were using mechanical traction by the mid-1970s and, since then, mechanization has advanced considerably. Demand is such today that the prices for hiring (or selling) equipment on the free market are three times as high as those of state agencies. The state, encouraging demand, is incapable of meeting the needs of small farmers, thus the hiring of machines and tractors by the richest fraction of the middle peasants is growing and leading to transfers of surplus within the private sector.

In Tunisia, over half the private farmers currently use mechanical equipment, but for every owner of equipment there are 16 hirers. The small peasants who use tractors do so almost exclusively on a hiring basis; of 200,000 small and middle farmers who use tractors, some 186,000 hire them from better-off peasants.

Social differentiation in the countryside also derives from the degree of openness to the market of the various categories of farms, which conditions the possibility of investing in the most profitable cash crop.

Generally speaking, cereal-growing, which is far and away the leading cash crop, is concentrated in the public sector and in small commodity or semi-commodity agriculture. Cereal prices are fixed by the state at levels that are not very rewarding compared to those of other agricultural products. Conversely, fruits and vegetables as well as livestock-raising are above all practised by medium-sized farms in the private sector. Their prices are free, and demand for them is rising strongly with the rise in average income in the towns.

In North Africa, and, to a lesser extent, in Egypt, agriculture is now organized into four productive systems differentiated by the size of farm, the labour process and the extent to which production is marketed.

The first productive system is that of smallholder property, that is, smallholdings generally situated in areas of low rainfall and no irrigation, using exclusively family labour on the basis of archaic work processes. Their essential function is to provide for the subsistence of the family group, which is why these smallholdings grow mainly food crops for subsistence. But the low yields of the land, the disguised unemployment and the need to buy consumer goods force those living on them to sell their labour and seek other opportunities for earning money. This productive system occupies only a small part of the agricultural land but a large proportion of the peasantry, generally more than one-third. It is linked to the market mainly through selling labour, although the smallholdings may market part of their production. It is mainly from here that the rural wage-earners are drawn, and also the rural unemployed; people who are likely to join the drift from the rural areas and emigrate.

The second productive system is that of the medium-sized commercial family property (five–20 hectares). This system has the distinctive feature of making occasional use of wage labour on areas large enough to enable more modern, often mechanized, working methods to be employed. Production is mainly for the market, and is more diversified than that of smallholdings (cereals, vegetables, forage crops). Nevertheless, yields remain rather low and arc subject to climatic conditions, and the incomes derived from farming are such that families need to seek other opportunities outside. This system weighs heavily on the peasantry and occupies a relatively large agricultural area.

The third productive system is that of upper-middle farms. Here, land leased may represent a not insignificant fraction of the area worked, but the areas owned and farmed directly generally make up the bulk of this system. Furthermore, the farm still remains a family holding, although it regularly uses wage labour. Production is exclusively intended for the market, working methods are always mechanized, cash crops more diversified, yields and earnings higher. There is no question here of selling one's labour outside the farm, nor of having recourse to hired equipment. Access to credit, and to irrigation are much easier. Because this productive system is able to adapt the structure of its production to the incentives of the price system, and because it enjoys implicit transfers of income at the expense of the majority sectors of the peasantry, it has been the most dynamic sector in Arab agriculture.

Finally, the fourth productive system is that of the large mechanized farm, based exclusively on wage labour and the marketing of production, whether it is of the capitalist or the state-run type.

Moroccan agrarian capitalism was able to adapt its production to the evolution of the market by developing fruit and vegetable production both for local demand and for export. In Tunisia, however, the large, private farm sector has shown little dynamism; the big landowners prefer to make money out of hiring equipment and extending areas (generally sown to cereals) rather than intensifying production. The yields of large Tunisian estates are very low: according to a survey conducted in 1980, the gross product per hectare of cereal-growing in the north of the country declines in inverse relation to the size of farm.

In Algeria, the state sector, which was established on the former settler lands, also has very low yields (some ten kilograms per hectare) despite the use of machines and fertilizers in the framework of large farms; its value added is not even enough to pay its workers' wages.

These four productive systems maintain unequal relations with one another, notably through the hiring of equipment and the permanent or occasional use of peasant wage workers. The maintenance and consolidation of family and smallholder property above all enables the other sectors of agriculture, and the national economy as a whole, to have a reserve of workers whose needs are in part taken care of by working the land.

With the consolidation of family and smallholder agriculture, use of peasant labour outside the farm is developing. It has been estimated that in Tunisia, over one-third of farmers currently devote over six months in the year to activities other than those needed by their own holding. The sale of labour takes place in the framework of family strategies to distribute the activities of members of the family between farming the plot it possesses, agricultural, and non-agricultural wage labour. As wage labour is less developed in agriculture, poor peasants seek a supplementary job in, especially, construction, industry and services. In Morocco, only 58% of the economically active rural population is employed by the agricultural sector. The development of non-agricultural employment in the countryside is even more advanced in Algeria, where half the economically active rural population is employed outside agriculture, mainly in construction and public works and services.

Growth, modernization and autonomy

There are many differences between Turkey and the Arab world, and within the latter between the open capitalist systems and the populist systems. These differences have to do with the history of each country and the international conjuncture. But none of the three systems has so far succeeded in ensuring a true modernization of any Mediterranean society, nor has any succeeded in overcoming a structural external imbalance.

Industrialization, unemployment and labour

Nowhere in the region has the development model applied succeeded in radically transforming the occupational structure of a growing population. Despite big differences between one country and another, industrialization has not led to the transfer of the majority of the economically active population to high productivity activities. The average level of social productivity remains extremely low, and only a minority of the labour force is integrated into modern forms of production and labour processes. This means that the undeniable economic development has not engendered a real modernization of society, if it is accepted that social behaviour and forms of collective thinking are largely conditioned by the mode of existence and labour.

Employment has expanded quite substantially during the last two decades, but this expansion has been inadequate given the demand for jobs and even more inadequate given the actually available labour force. This latter is in fact much larger than the number of job seekers since it includes a large number of potential workers who, for social and cultural reasons, do not enter the labour market. This is the case with the vast majority of women and of some of the young people who have had no schooling, and also of a significant number of disguised unemployed vaguely occupied on the fringes of the urban informal sector.

In Turkey, the economically active population rose from 13 to 19 million between 1965 and 1980, but the unemployment rate is still currently around 20% of the economically active population, and over half the population still live in the countryside. In Tunisia, employment rose over the same period from 900,000 to 1.6 million but the unemployment rate remained stable, at around 15%, for men, who make up the vast majority of the economically active population. But, proportionately, female employment has increased much more.

All over the region, due to the destruction of old social relations, changes in family structure, monetarization of the traditional economy and the diffusion of new patterns of consumption, a growing number of women are looking for work. Conversely, the spread of education among women has led, notably among graduate middle-class women, to the adoption of new cultural models in which a positive integration into the family and society now rests on a skilled vocational activity. In short, whether it is based on material need or results from new social norms, women are increasingly entering into the labour market. Government policy does not create women's demand for jobs, but it can help or hinder the extent to which this demand is met by the country's economy. The argument sometimes put forward that women occupy jobs at the expense of men is not only absurd, in that it amounts to saying that the employment of a few workers creates unemployment for others, but tendentious, because it assumes that women permanently have an inferior status in society. In Tunisia, 13% of the female population of working age are in the labour market, and a little over one-quarter of the economically active population is made up of women. These proportions, though far from those to be found in Europe, America or Asia, make Tunisia, in this respect, the most advanced country in the Arab world. Such an outcome would not have been possible without action by the state in favour of helping the economy to meet women's demand for jobs.

In Morocco, the official unemployment rate was 15% in 1980, and in the same year almost 300,000 unemployed were officially counted in the urban centres, but to them must be added some 500,000 rural unemployed. In Algeria, the number of unemployed long remained fixed at the level of 700,000, but it has increased in recent years. In Egypt, one-fifth of an economically active population estimated at about 10 million people is officially unemployed. But official estimates of unemployment seriously underestimate the scale of real underemployment. Not only because the vast majority of women and young people excluded from schooling are not counted as job-seekers, but also because a significant fraction of the population that, theoretically, is employed is in reality largely under-employed, in agriculture and rural and urban services. Anyone familiar with the cities of the Arab world will find the official figures barely credible, knowing the vast mass of jobless, mostly young men, who hang around in the urban areas at times when factories, offices and businesses are normally open.

The pressure that the demand for jobs (or the supply of labour) puts on the economic system is, of course, greatly increased by a very high population growth rate. Moreover, historically, the employment growth rates observed in Turkey and the Arab world are far from insignificant if they are compared with those that Europe experienced at the time of its industrialization. But the demographic constraints are also an effect of the inability of the development model to modernize society.

Until recent years, emigration helped to make the gap between the supply and demand for jobs less explosive. There are estimated to be three million Turks abroad, a figure equal to the number of persons employed in Turkish industry or one-sixth of the economically active population. There are also three million Egyptian emigrants, who have settled in Arab oil-producing countries rather than in Europe, equal to one-third of the economically active population in Egypt. In North Africa, where the emigration of workers, mainly to Europe, is older, emigrants also account for a very high proportion of the labour force. But the crisis in Europe since the mid-1970s and the crisis in the oil-producing countries since the mid-1980s have removed this safety valve just when economic difficulties were worsening in Turkey, Egypt and the Maghreb and leading to an increase in unemployment.

The exclusion of social strata involving varying numbers of people from the benefits of economic growth results from the inability of states to institute the most appropriate procedures for raising social labour productivity. But it also derives from the gap between the rhythm of social modernization and the real or potential sources available to society. Urbanization and changes in lifestyle and consumption patterns are too rapid in relation to the objective possibilites of increasing agricultural and industrial production, although it is true that these are far from having been fully exploited. Such a gap is obviously more visible in the open capitalism countries in which a much more unequal distribution of incomes than elsewhere implies a greater waste of scarce resources. But whether social modernization affects minorities, as is the case in these countries, or extends to larger fractions of the population, as in Egypt or Algeria, it continues to be defined by access to modern goods and services and the development of individual consumption. In the framework of the current world price system, however, this relative and limited modernization has been obtained only thanks to the availability of various rents, rather than through a mobilization of the productive capacities of society. With the crisis resulting from the reduction of these rents, it has become clearer that the social spread of this modernization will have to be increasingly restricted. While the crisis underlines the urgency of developing social productivity, it also calls for a questioning of the societies concerned, and their elites, as to their condition and their future, since clearly, the aspiration to spread Western levels of living standards and forms of consumption to the whole population is unrealistic.

Such sparsely inhabited oil-producing countries, as Saudi Arabia, Libya, Iraq and the Gulf countries, have not experienced a labour surplus. On the contrary, the very high growth fuelled by oil rent has led to an acute shortage of workers made up for by a massive recourse to immigration. But the vast majority of women continue to be excluded from the economic system.

In fact, the inability of the economic model to absorb the total available labour force is a result of the inability of the industrial sector to take on a significant fraction of the population.

The 'formal tertiarization' of labour

In Turkey, as in the Arab countries, industry is unable, despite relatively high growth rates, significantly to increase the number of people it employs and continues to involve a relatively limited social area.

The share of industry in the distribution of total employment in Turkey rose from 10% in 1965 to 16% in 1982, and agriculture continues to occupy over half the economically active population. In Egypt, the share of industrial employment increased only slightly during the same period, from 12% to 19%, but services now occupy over 40% of the population.

The Maghreb countries have moved in more or less the same direction. In Tunisia, industry has shown a greater potential for job creation than elsewhere, but it still absorbs only a minority of the population. Industry in fact provided 29% of the jobs created outside agriculture in the 1960s and 40% in the 1970s but it still occupies only one-fifth of the labour force. Agricultural and rural migration has grown markedly faster in Turkey and Egypt, but of the workers driven out of agriculture more have been absorbed by services and construction than by industry. In Morocco, industrial employment has increased far less and agriculture undoubtedly remains the principal sector of employment, since it still accounts for 60% of employment. Craft activities are very important here, having been able to adapt to new conditions by diversifying their products and customers, from the production of practical objects for people in the lowest urban stratum excluded from the modern market, to the manufacture of art objects, by way of small-scale service activities associated with household equipment and tourism, the purchasing power of the majority of the population fell. But the processing industry created only 45,000 jobs between 1960 and 1978. Furthermore, Moroccan industry is built mainly on a large number of small workshops employing small numbers of people. The option in favour of light industry has not, as it has in Tunisia, been synonymous with job creation. In Algeria, agriculture now occupies only one-quarter of the economically active population, a much lower proportion than in other Arab countries and in Turkey, but industrial employment accounts for only 14% of total employment, despite a vast increase in its numbers, which have risen from 116,000 to 514,000 since the 1960s. Of every ten jobs created, four have been in the administration and services, three in construction and only two in industry. This particular profile of job creation does not correspond to the historical experience of the industrialized countries in either the East or the West; in those countries, economic development was accompanied by a transfer of the economically active population from the primary to the secondary sector, then from the latter to the tertiary sector. In any case, it reflects, above all in the populist model countries but also in the oil-producing countries, the relative advance of the modernization of the environment and society over industrialization. Growth, like job creation, rests more on the development of infrastructure (construction, public works, communications, distribution) and the development of education, health and administrative services than on the expansion of productive activities, in agriculture and industry.

Although the Gulf countries, Iraq and Libya are faced with a shortage of labour, the structure of their economically active population is also marked by a pronounced premature 'tertiarization'. As early as the 1960s, the low level of their agricultural development was reflected in the fact that construction and services headed the list in the distribution of employment: in Saudi Arabia, for example, only 31% of the economically active population were engaged in agriculture in 1966 (but 15% in herding). And since then, the fall in agricultural employment (24% in 1980) and especially herding (4%) has benefited services (55% in 1980) and construction far more than it has industry, whose share remains minimal, at 4%. Saudi society has became a rentier society, whose activities rest on the distribution of petroleum rent by the state.

Growth – and this is another feature common to Turkey and the Arab states – is occurring in the context of a permanent trade deficit. All non oil-producing countries have experienced a deficit on their external trade over the last three decades, and in the oil-producing countries it is simply masked by the rent extracted from hydrocarbons.

Growth and external deficit

In Turkey, the export cover rate has long been on a downward trend, falling from 70% in the 1960s to 36% by the end of the 1970s, while the ratio of imports to GNP rose from 7% to 11%. Industrialization has not succeeded in reducing the economy's global dependence on external trade. The deterioration in the terms of trade has accentuated the combined effect of the increase in the volume of imports and the stagnation of exports. In Egypt, the cover rate fell from 66% in 1965 to 36% in 1975, and 33% in 1982. It was 66% in Tunisia in 1981 after having fallen to lower figures a few years earlier. In Morocco, it has varied between 40% and 60% since the mid-1970s. Financing of the trade deficit has been provided either by official and private external capital flows or by migrants' remittances and tourism.

In Turkey the inflow of foreign capital made up for virtually the entire deficit in the 1950s; this capital consisted principally of US official transfers and World Bank and IMF credits. Turkey's highly strategic position, and the fact that it has deliberately opted for the Western camp, obviously explains why official US agencies covered the trade deficit, although it is also true that this aid led to flows of 'made in USA' goods to Turkey and helped to keep this deficit going. But during the 1960s and 1970s, the proportion of US public funds fell, while that of bank loans and funds repatriated by migrants rose. Initially, the Egyptian deficit was similarly covered by foreign aid which, during the Nasser period, represented on average a quarter of imports from the USSR and USA. Later, external aid, mainly from the Gulf countries, continued to play a significant role. After the signing of the Camp David agreements, US aid once again became important.

The inflow of external capital also played a major role in making up for the trade deficit throughout the period in Morocco and Tunisia. In the Arab countries and Turkey, official external aid was gradually reduced as private bank credits and direct investment rose, but the replacement of government transfers by private funds was most noticeable in places where the level of industrialization was highest. Thus, in Tunisia, between 1971 and 1981, the proportion of private transfers rose from 36% to 69% of external capital flows, while in Morocco, public credits remain markedly more important.

As external indebtedness rose, however, a growing proportion of external capital flows went to service the debt, but also to pay profits and dues associated with direct investment. Capital inflows were thus less and less able to finance the commercial deficit.

But this commercial deficit was also covered in part, notably from the 1970s onwards, by other sources of income, such as remittances by migrants and earnings from tourism. At the beginning of the 1980s, funds repatriated by migrant workers represented 17% of the value of imports in Tunisia, 44% in Morocco, 55% in Turkey and 61% in Egypt. To that must be added earnings from tourism, which are very important in most Mediterranean countries, and dues collected for transit and navigation, such as for use of the Suez Canal. Finally, it may be argued that the rent enjoyed by oil- and gas-exporting countries makes it possible to hide a position that is in deficit in real terms. This is, of course, the situation of such big producers as Libya, Arabia and the Gulf emirates, but it is also the case elsewhere. In Syria, oil exports rose to over one billion dollars in 1979, equal to the amount of Arab aid received in that year. In Egypt and Tunisia, they now cover a significant fraction of external purchases.

Such heavy dependence of the countries of the region on external capital, and various rents from exports of oil or labour or from tourism, does not mean that these countries do not suffer transfers of value to the benefit of the more developed countries. After all, the commercial deficit is also the result of the mechanisms of unequal exchange and the under-valuation of goods exported, whether primary products (except for oil and gas) or bottom-of-the-range labour-intensive industrial goods. And the balance of payments deficit arises in part from payments made to external operators in the form of dues, superprofits and interest.

Whatever the case, the persistence of the commercial deficit and the balance of payments deficit means that neither in Turkey nor in the Arab world has there been any success in adjusting labour to expenditure, resources to needs, in the framework of the current system of world prices. A relatively high proportion of consumption and investment is financed by incomes that do not reflect productive activity in the country. In other words, reproduction of the economic and social conditions of development depends on the availability of revenues over which the societies concerned have, by definition, no control.

The reduction of public and private Western credits and remittances by migrants, from the mid-1970s on, as a result of the recession in Europe and North America logically precipitated the crisis of development models, at least in non oil-producing countries, but the oil-exporting countries were to experience the same fate a few years later, when oil prices on international markets fell and the dollar declined in value.

The crisis and the responses to the crisis

In Turkey, the rise in real wages was not assimilated by an improvement in labour productivity, since it was made possible by the existence of rents from protection and the exchange rate in industry. For the private industrial sector, protectionism made it possible to fix domestic prices at a level such that the gap between them and international prices was markedly higher than the gap between local and foreign production costs. In addition, the overvaluation of the currency represented a disguised subsidy to imports of capital goods, semi-finished products, energy and raw materials. The public sector, whose specialization went hand-in-hand with the reduction in the relative price of its products, benefited very little from rents from protection and the exchange rate, but it enjoyed a quasi-rent from its position, since its deficits were regularly covered by the Treasury, that is, by people's savings and by printing money. In the last analysis, these various rents rested on the possibility of a permanent recourse to imports without a corresponding expansion of exports. But the reduction in credits and remittances from migrants led to a sharp fall in import capacity just when the oil price rise was imposing further burdens on the trade balance. After 1975, the level of imports thus tended to stagnate, signalling the crisis in the model of growth and income distribution. The slowdown of expansion following the reduction of external rents meant that the system could no longer absorb rising peasant wages and incomes without endangering the profit level. By the late 1970s, the percentage of wages in industrial value added had climbed to over 37% as against 30% in 1975 and profits had declined for the first time since the 1950s. In 1977, the effects of the crisis were already considerable, with foreign exchange reserves almost exhausted, a large budget and balance of payments deficit and an external debt rising to 13 billion dollars. Inflation, which had remained below 10% until the beginning of the 1970s, passed 20% in 1976 and 100% two years later. It was in this context that the IMF, whose green light is now a condition for obtaining new credits from Western banks, intervened twice, in 1977 and 1980, to demand the implementation of the usual adjustment measures. The Turkish pound was devalued, and a system of flexible foreign exchange rates was introduced, which accentuated the depreciation of the national currency. All price controls were eliminated as well as the subsidy on public sector products. Interest rates were raised sharply and the agencies controlling external trade were dismantled. The agricultural price support policy was abandoned and wages were subject to a harsh freeze.

The crisis and the implementation of the IMF programmes inaugurated a period of serious social and political instability that culminated with the intervention of the army, which felt itself to be the guardian of the Kemalist project of national capitalist development and social modernization. But the army, which had intervened in 1960 against an economic policy judged to be too anti-*étatiste*, put into effect an ultra-liberal policy in an attempt to overcome the crisis. It took up the whole IMF package of measures, freed prices, banned strikes and suspended collective bargaining. During the first three years of its application, this policy seemed to be achieving some success. Inflation was brought down from over 100% in 1980 to 29% in 1983, and exports, particularly industrial exports, rose strongly, while imports were stabilized. Exports of manufactured goods, which accounted for a quarter of total exports in 1975, were two-fifths in 1982. The deficit in the trade balance was reduced to an amount slightly higher than that of migrants' remittances, about two billion dollars per annum. The country, which was almost on the point of ceasing payments in 1980, recovered its capacity to repay the external debt. Finally, growth, which had run out of steam in the late 1970s, resumed, reaching rates of 5.5% in 1984 and 8% in 1986.

And yet, in 1984 inflation rose again to 52% and remained high. Imports, especially of capital goods, began to rise again despite the continuous devaluation of the Turkish pound, while exports suffered from the contraction of markets in the Middle East. The commercial balance went into growing deficit again, aggravated by the decline in the remittances of migrants and earnings from tourism, and debt service reached 51% of export receipts.

Inflows of external capital remained modest, despite the fact that the government stepped up its incentives and guarantees. Finally, the dear credit policy precipitated the bankruptcy of numerous companies and discouraged productive investment. As for the privatization of the public sector, it has had no more success than it did in the 1950s. The best known cases are those of the Euphrates dam company and the Bosphorus bridge company.

The effects of the supply-side policy inspired by Reaganism and the IMF are visibly running out. And the government is now seeking to temper its liberal creed, by coming to the help of threatened companies, increasing prices in the public sector, relaxing control of the money supply, and subsidizing some export products. Meanwhile, this policy has had an immense social cost. Unemployment has worsened considerably and, according to official figures, now affects over 20% of the economically active population or almost four million people. On a base of 100 in 1974, the real wage index fell from 176 in 1978 to 87 in 1982.

In recent years, Turkey has provided the most extreme example of the opening-up of a state capitalist system to both economic liberalism and international competition. State policies have been radically reoriented, and it is not accidental that the shift coincided with the imposition of a military regime. In the Arab countries, the development model's inherent blockages became apparent earlier, at the beginning of the 1970s, as in Egypt and Tunisia, or much later, in the mid-1980s, as in the oil-producing countries.

The Egyptian *infitah* ('open door') policy put into effect in 1974, was in fact determined more by the change in the relations of force at the top of the state, following Nasser's death, than by economic difficulties. But these latter were very real, as the agrarian reform and industrialization proved unable to prevent the swelling of urban unemployment and the aggravation of the external deficit consequent upon Nasser's social policy. The *infitah* appears as the restoration of capitalism and opening up to the international market, with a view to increasing the efficiency of the economic system and restoring the balance between productivity and income. Contrary to what was to happen later in Turkey, liberalization of the economy began in Egypt under the auspices of a liberalization of political life, which, although limited, was real, and it was not called into question despite the serious tensions the country was to experience. Land transactions were deregulated, as the ceiling established by the agrarian reform on landownership was lifted and land developed by the state was sold to private individuals. The government promoted greater autonomy for public enterprises and the transformation of some of them into joint-stock companies, which could be sold to private investors. The areas in which external capital could operate were considerably extended and it was granted a special legal regime, particularly in the area of labour legislation and foreign exchange control. The state monopoly on banking was abolished and the creation of private banks, both Egyptian and foreign, was authorized. The state abandoned its monopoly on importation and established a system of imports without official transfers, which made it possible to use foreign exchange in the possession of private individuals for imports. Finally, foreign exchange controls were abolished and the Egyptian pound was devalued. This policy, turning its back completely on the Nasserite system, thus aimed at removing the obstacles of the concentration of landownership, privatizing a large part of the public sector, attracting foreign investors and, finally, removing state control over banking, the currency and external trade. At the same time, a systematic attempt was made to replace administrative procedures for fixing prices and income distribution by market mechanisms, that is, to eliminate the subsidies on food products (which represented 10–15% of GNP) and to introduce realistic pricing into public enterprises.

The Egyptian *infitah* policy, however, is far from having achieved its declared objectives of raising the productivity of the economy and reducing the external deficit. By the beginning of the 1980s, no more than a few hundred enterprises had been created in the framework of the new provisions and their cumulative capital was relatively extremely modest. As in Turkey, the private sector has been unable or unwilling to take over public enterprises. In addition, the bulk of national private investment has gone into non-industrial activities, offering better profit possibilities and fewer risks. As for the private capital that has been invested in industry, it has kept to the traditional areas of food, plastics and building materials. This is thus a far cry from the expected rush of aggressive entrepreneurs anxious to push aside the barriers raised by the bureaucracy to the unfettered deployment of their management capacities in industry.

Foreign capital has not rushed either to replace the state or create new enterprises, although a few investors have been attracted by the concessions on offer. It has been calculated that 64% of the financing of new private investment has been made by Egyptian capital, 19% by Arab capital and only 17% by Western capital. Here, too, the expected inflow of foreign capital has not materialized. Neither local capital nor foreign capital has sought to occupy the commanding heights of the Egyptian productive apparatus so as to orient its activity in accordance with world price relations. Moreover, foreign investment has not been directed into export activities, preferring the domestic market. Thus it will come as no surprise to learn that public investment continues to account for the bulk of investment, although it is true that the share of private capital in total investment rose form 9% to 15%.

Realistic pricing, too, soon met its limits in the resistance of the popular urban strata. The state raised prices only gradually and indirectly and even international financial agencies finally accepted the maintenance of subsidies for basic essentials in order to maintain social and political stability.

Conversely, the 'open door' policy has led to an unprecedented expansion of unproductive activities and speculation. Banking activity has boomed, as imports have increased. More serious is the fact that it has led to a worsening of the crisis of the state-owned productive system, in particular in industry. The state industrial sector is suffering from the emigration of cadres and skilled workers, and shortages of spare parts and intermediate goods, just when it is facing foreign competition. The *infitah* is seeing a process of relative de-industrialization since, during the 1980s industry's share in investment was only 12.5%. Its contribution to GNP is tending to stabilize and its capacity to mop up unemployment to fall even further: it created only 30,000 jobs per annum between 1975 and 1980, against a demand for jobs at least ten times higher.

The powerlessness of the 'open door' policy to transform the Egyptian economic system in the direction of an export-oriented industrial capitalism is now amply established. Rather than a reorientation of growth in the framework of greater integration into the international division of labour, this policy has led to a speculative economy based on the recycling of the rents derived from oil, on external aid, the running of the Suez canal and remittances from migrants in the Gulf. Unlike the Turkish experience, the policy of *infitah* in Egypt has thus not succeeded in rationalizing the economic and social system from the viewpoint of local and foreign capitalism. It is a moot point whether the unexpected expansion of oil production and migrants' remittances in the late 1970s has not removed the need to produce exportable industrial surpluses and diverted private investment away from industry by opening up vast opportunities for speculative gains. But, on the other hand, emigrants' remittances have attenuated the social tensions engendered by the 'open door' policy, by limiting the effects of the fall in real wages and the extension of unemployment on the level of popular consumption. It is any case undeniable that while the *infitah* has not led to a new industrial specialization, it has very decidedly benefited Egypt's reintegration into the Western economic and political system. The United States and Western Europe have become the country's main commercial and financial partners. Egypt signed a convention with the EEC in 1976, and in 1978 signed an agreement with the IMF that made the granting of credits conditional on the implementation of structural reforms, which in any event were on the government's policy agenda.

Without being as clearly articulated, as in Egypt or Turkey, the 'open door' policy has also been on the agenda in the Maghreb, since the early 1970s in Tunisia, a little later in Morocco and since the 1980s in Algeria. Everywhere, the state is seeking to develop industrial exports and attract foreign capital, in order to reduce the external deficit and stimulate the expansion of employment. In Tunisia, a law was adopted in 1972, whose objective was to attract foreign investors and encourage export industry by granting a series of tax concessions and exemptions in the areas of the import of components and semi-finished goods, labour legislation and exchange control regulations. The result was a tripling of industrial exports during the 1970s. Clothing and the assembly of electric and electronic products on a labour-intensive basis make up the bulk of the enterprises created in the framework of the provisions of this law. These, then, are investments based on the low level of wages relative to labour productivity. Available estimates suggest that the level of labour productivity in these activities is almost comparable to that in European factories, whereas the gap in wages is of the order of 1:7 in clothing and 1:4 in the electrical and mechanical engineering industries. In addition, while Tunisian private capital participated in investment, most of the firms formed were foreign. Their cumulative investments, however, represented only 6% of industrial investment between 1972 and 1981 and 17% of the number of industrial jobs created. The receipts derived from the export of industrial goods covered approximately one-eighth of the commercial deficit. But the export industry is an enclave in the country's economy and is, in fact, more like a free zone. A survey conducted by the OECD shows that the export enterprises buy almost nothing in the country, do their repairs in Europe, use obsolete, secondhand equipment and employ unskilled staff. Finally, they have no autonomy in product planning and marketing, as most of them are involved in subcontracting.

Moroccan state policy began to favour industrial exports from the mid-1970s, by giving priority to the chemical industries associated with the processing of phosphate, and international subcontracting. About half the firms in the mechanical engineering and electrical industries are involved in subcontracting, but four-fifths of industrial exports come from the foodstuffs industry, textiles and chemicals. Over a quarter of textile production was being exported in the late 1970s, but export activities still accounted for only one-tenth of overall industrial production. There, too, local integration of export industry is very slight and acquisition of know-how very limited.

This precocious open door policy pursued in Tunisia and Morocco did not contribute anything to reducing the external deficit, nor did it improve the economy's performance in the areas of growth, investment or employment. In the early 1980s, exports covered only around 60% of imports, and successive devaluations of the Tunisian dinar and Moroccan dirham have had no noticeable impact on industrial exports. In 1986, Tunisia's external debt exceeded seven billion dollars, and debt servicing was equal to almost 40% of export earnings, while Morocco's external debt exceeded 14 billion dollars and constituted a burden that had become intolerable. During the 1980s, the two countries have seen their growth and investment rates fall, and this has naturally led to a reduction in job creation and a worsening of unemployment. Despite everything, government policies continue to seek a greater opening up to capitalism and the international division of labour, with the illusion of being able to reduce external indebtedness. But pressures, instructions even, from the creditor countries, the banks and the IMF, leave states with little room for manoeuvre, at least in the short term. In both Morocco and Tunisia, current policies are moving towards the complete freeing of prices, import liberalization, currency devaluation and privatization of the public sector. The objective is to bring down real wages so as to reduce popular consumption and production costs in industry and promote both exports and foreign investment. In fact, in the 1980s, with the reduction of subsidies on basic essentials, the freezing of wages and the worsening of unemployment and inflation, but without leading to any marked growth in industrial exports or attracting foreign investors.

In Algeria, the crisis brought about by the fall in the dollar and in the price of oil over the last two years has shown the vulnerability of the economy to foreign factors and accentuated the constraints they impose. In 1986, the balance of trade was in deficit for the first time in ten years and export receipts fell by 40%. Since then a policy of adjustment has been applied, consisting essentially in a reduction in investment rather than in consumption. Consumption is nonetheless affected, in that state imports have been cut back and that the fall in investment has led to worsening unemployment. Rationalization of the public sector is also on the agenda, notably through the autonomization of state enterprises which are henceforth required to balance their books. Also, state sector lands have been distributed to agricultural workers. The permissible forms of association between public sector firms and foreign companies have been broadened, and the tourist industry has been thrown open to foreign investment. For the moment, there is no question of breaking the state monopoly on banking, credit, foreign trade, hydrocarbons and the major industrial installations, or of encouraging foreign investments in any sectors except tourism. Private capital is encouraged to invest but only in the context of sub-contracting from the big state enterprises. By analogy with the eastern bloc countries, the policy of structural adjustment aims at developing competition as an alternative to economic planning, rather than the privatization of the economy. But if the exapansion of local and foreign capitalism seems to be muzzled, the commoditization of society is developing through the partial privatization of the health system, the creation of a property market with the sale of state property to their tenants and the freeing of property dealing, the authorization given to individuals in possession of foreign exchange to import private vehicles, and machines for their businesses, and the introduction in banks of foreign exchange accounts held by entrepreneurs or even ordinary individuals.

No more than in Turkey has the open door policy in the Arab world made the economy advance in the direction of higher growth, a less unequal distribution of income, greater job creation and less external dependence. Expansion is compromised by the fall in investment as a result of the adjustment policy, the number of jobs is growing only very slowly and the development of exports does not even appear, as in Turkey, to offer serious, real possibilities of having a knock-on effect on other sectors of the economy. On the contrary, average income is falling and income distribution is becoming increasingly unequal. As for external autonomy, there is less prospect of that than ever. Clearly, the greater openness to capitalism and the international division of labour is unlikely to facilitate overcoming the blockages of the state capitalism model and that an alternative path becomes more urgent with each passing day.

Index